Adapting to Sea Level Rise in the Coastal Zone

Law and Policy Considerations

Adapting to Sea Level Rise in the Coastal Zone

Law and Policy Considerations

Chad J. McGuire

Routledge
Taylor & Francis Group
LONDON AND NEW YORK

First published 2013 by Taylor & Francis Group

2 Park Square, Milton Park, Abingdon, Oxfordshire OX14 4RN
711 Third Avenue, New York, NY 10017

Routledge is an imprint of the Taylor & Francis Group, an informa business

First issued in paperback 2018

ISBN-13: 978-1-4665-5980-6 (hbk)
ISBN-13: 978-1-138-37995-4 (pbk)

Library of Congress Cataloging-in-Publication Data

McGuire, Chad J.
Adapting to sea level rise in the coastal zone : law and policy considerations / Chad J. McGuire.
pages cm.
Includes bibliographical references and index.
ISBN 978-1-4665-5980-6 (hardcover : alk. paper)
1. Sea level--Environmental aspects. 2. Coast changes. 3. Coastal zone management--Government policy. 4. Environmental policy. 5. Environmental law. I. Title.

GC89.M373 2013
363.34--dc23 2013003038

Contents

About the Author

Chad J. McGuire is a professor of environmental policy within the Department of Public Policy at the University of Massachusetts, Dartmouth. He is also an attorney licensed to practice law in California, Nevada, and Massachusetts. His background is in environmental law and environmental science. Chad teaches, writes, and practices in the fields of environmental law, policy, sustainability, and dispute resolution. He has published more than 20 scholarly and applied articles on these topics over the past 10 years. He has worked on policy issues related to fisheries management, climate change, coastal management, and land use patterns. His expertise has been sought in both private and public forums, and he has served on committees for both non-profit and government entities. He has more than 15 years of experience in the environmental management field. This book is based on his work in studying the problems that arise at the intersection of law and policy, in this case those problems that are directly related to coastal management planning necessitated by sea level rise.

In addition to this text, Professor McGuire is also the author of *Environmental Decision-Making in Context: A Toolbox*, published in 2012 by CRC Press, and part of the American Society for Public Administration's Series in Public Administration and Public Policy.

Chapter 1

Introduction

The coastal zone is a dynamic place, and this statement has been true throughout recorded human history. For as long as humans have been inhabiting coastal areas and recording what occurs in their environments, coastal zones have been defined through dynamic interactions. The area between land and sea shifts as the tides move landward and seaward. Coastal storms arrive at the shore, moving large amounts of sand in the process, sometimes eroding coastal areas and other times adding dry landmass to the coast. Humans have also influenced the contours of coastal areas through dredging and filling operations, intensifying these operations with the advent of technological innovations throughout history. More recently, humans have begun to defend against tide and sea by building walls and other barriers that attempt to limit the approach of the ocean into the land.

All the phenomena mentioned above reinforce the changing nature of coastal areas; as such, it should come as little surprise that coastal boundaries do not remain fixed in time and place but rather are in a state of constant flux. This text is not about the kinds of changes mentioned above; changes that have been observed as part of the *background conditions* of coastal areas. Rather, this text is focused on a more recent development: observed sea level rise. Over the past several decades, sea level rise has become a relatively constant phenomenon in most coastal regions around the globe. Scientific inquiry into the causes of sea level rise has placed human activities, specifically carbon emissions into our atmosphere, as a major cause of this observed phenomenon. By increasing the amount of carbon in our atmosphere, the Earth is warming and thus causing water held as ice to begin melting. As the water melts, it runs off into the Earth's oceans, causing sea level rise. In addition, warmer temperatures cause water molecules to heat up, leading to the thermal expansion of water. This thermal expansion of water in our oceans is also adding to observed sea level rise.

Current sea level rise is different from the other phenomena mentioned above that cause changes to coastal areas. This currently observed sea level rise is a constant process that is occurring and, based on current best scientific estimates, will continue to occur for our foreseeable future. The question presented today is really neither about whether sea level rise is happening nor whether it will continue into the future, but rather how far sea levels will rise. This is an important question because the future extent of sea level rise defines what actions must be taken today in order to deal with the problem. For example, if the sea level only rises a few inches over the next 100 years, then there is little need to change current uses or expectations in most coastal regions around the globe. However, if sea level rises substantially—say 20 feet—over the next 100 years, then planning must occur immediately to alter the uses and expectations of coastal areas that will be impacted by such a rise in sea levels.

The current state of science contains a lot of uncertainty about how far sea levels will rise over the coming decades, and this uncertainty is problematic when considering planning options. There certainly is enough ice on Earth to raise sea levels by 20 feet (there is enough ice to raise average sea levels over 200 feet and more, which would be devastating). The question is not what is possible, but rather what is probable when contemplating planning for adaptation toward sea level rise. By knowing what is probable in terms of sea level rise, planners are better able to make decisions about how to plan for this anticipated rise. When information about future sea level rise is certain, the choices about what needs to be done today are clear. When the information about future sea level rise is uncertain, the choices about what should be done today are less clear. This brings us to the purpose of this text: to explain the causes of sea level rise, the policy options related to sea level rise planning, and finally to consider the legal implications of different policy directions.

It is important to note that this book is not about preventing sea level rise or mitigating its impacts. Rather, this text focuses on sea level rise as a given and discusses policy planning issues that arise when focusing solely on *adaptation* strategies to sea level rise. Thus, the key difficulty presented in this text surrounds making policy choices about how to adapt to *future* sea level rise when there is *uncertainty* about how high sea levels will rise in the future. If we knew the extent of sea level rise in the future, then it would be much easier to take action today with planning recommendations that limit the amount of harm caused to coastal areas in the future; for example, knowing sea level rise will inundate a coastal town in 10 years provides enough certainty to justify a policy that requires the retreat of that area within a 10-year period. However, the high degree of uncertainty over the extent of future sea level rise makes the task of planning today difficult. Not only is the task difficult in terms of anticipating future unknown events, but certain laws—including judicial interpretations of those laws—create even greater difficulties in limiting how coastal land is used without triggering a requirement that government pay citizens for planning that prohibits current uses of private land.

The goal of this text is to distill the science, policy, and legal issues surrounding sea level rise into a coherent set of principles that both identify important issues for consideration, and then discuss those issues in light of current law and policy paradigms. By understanding how the current state of the law impacts policy choices, the hope is to provide those who are involved in studying or practicing coastal management a set of analytical tools to better understand the relationships between policy choices and legal frameworks impacting those choices. As shown in the text, the law often provides limitations on government actions, particularly foundational constitutional principles that cannot be easily overcome by simply passing a law or regulation that violates these foundational principles. Thus, understanding these legal limits on public planning for sea level rise is an important starting point in making informed policy choices. Once these limitations are understood, policy directions can be better mapped out. Sometimes the law will limit what seems like a superior set of policy choices for adaptation by making the decision untenable because of the financial consequences that might apply to such policy choices, as is the case with the Fifth Amendment of the United States Constitution's requirement that just compensation be paid for the taking of private property. Thus, policy directions that amount to a "taking" of private property, whether directly or indirectly, may require that the government pay full freight for the policy direction, and this legal requirement may make the policy choice inferior because of its costs even though it is superior in other measures. This is but one example of how legal frameworks can impact policy directions. The different impacts legal frameworks can have on policy choices are highlighted in greater detail throughout this text.

This book has been divided into three main sections, each section divided into its own chapter, to clearly accomplish the goal identified above. The first section, Chapter 2, focuses on the science behind sea level rise. The goal of this chapter is to connect currently observed sea level rise with the science behind future predictions of how far sea level will rise, and also to provide the reader with some basic understanding of how different coastal features can influence the impact of sea level rise. Collectively, this information is meant to provide a foundation of knowledge about sea level rise dynamics, including the forces involved in its causes, as well as the forces and features that influence its impact on coastal areas. At the conclusion of Chapter 2, the reader should have a solid foundation of the basic factors that influence the processes and impacts of sea level rise on coastal areas. Armed with this information, the reader is ready to place policy choices into context.

Chapter 3 focuses on the policy options that are available in adapting to sea level rise. It begins with a section on understanding basic policy goals by discussing *net benefit* quantification methods (a way of rank ordering different adaptation approaches), as well as understanding the role and impact of *uncertainty* in developing and choosing between sea level rise policy approaches. After discussing these concepts, Chapter 3 focuses on the different kinds of adaptation approaches that may be taken in dealing with sea level rise. Three main categories of approaches are highlighted: (1) staying at the shore, (2) retreating from the shore, and (3) a mix of hybrid approaches that employs some measure of stay and retreat options. Finally, Chapter 3 concludes by considering adaptation approaches from economic, social, and ecological viewpoints. Viewing approaches through these different lenses can alter the ways in which benefits and costs are calculated, thus altering a net benefit analysis. Understanding how different lenses can influence a determination of net benefits is an important part of identifying and choosing between policy directions even without considering the impact legal frameworks have on choosing between competing adaptation choices.

Chapter 4 summarizes the major legal considerations that influence adaptation choices toward sea level rise. Critical to this discussion is understanding the nature of property rights in land generally, and how these rights are viewed in combination with constitutional principles that protect private property rights (Fifth Amendment protections) and constitutional principles that empower state and local governments to protect citizens from known and reasonably understood dangers (Tenth Amendment police powers). The mix of constitutional powers and prohibitions provides the context from which policy planning that impacts private property rights begins; any policy that influences private property rights must be understood in relation to the powers and limitations placed on government in regulating land. This discussion of constitutional power interactions between the Fifth and Tenth Amendments to the U.S. Constitution provides a grounding point for this chapter. From this grounding point, the chapter discusses the legal issues from the *role* government takes in implementing policy directions. For example, when government acts in its regulatory capacity to develop and implement sea level rise adaptation policies, there is a greater chance that regulation may trigger a *regulatory taking* of private property rights when the policy results in limitations placed on the use of private property rights. Contrarily, if government adopts a policy direction that stems from a nonregulatory stance, then there is less chance the policy will implicate Fifth Amendment prohibitions on the taking of private property. The ways in which government may act outside its regulatory nature are discussed in some depth in this chapter offering policy makers the opportunity to understand how to develop adaptation policies that stand a better chance of avoiding legal frameworks like taking challenges.

Collectively, the three main chapters identified above are meant to provide a summary understanding of the law and policy considerations that impact coastal zone planning for sea level rise adaptation strategies. Each chapter introduces its major goals and, where appropriate, links those goals to the information provided in other chapters within the text. Although the text is meant to flow from start to finish, the chapters can be read out of sequence as references are made to other sections of text where appropriate. Because the text discusses three related disciplines (science, policy, and law), which each require significant depth for full understanding, the text does not serve as a comprehensive discussion on any one topic. The lack of depth in any one topic is necessitated in order to provide an understanding of the interactions between the science, law, and policy issues discussed. However, where appropriate, citations to primary materials with greater emphasis in each area are provided to allow the reader to gain additional insight and understanding where they deem appropriate. With this limitation stated, the hope is that each reader will find important conceptual connections between the kinds of actions needed to be taken in coastal planning to help proactively deal with sea level rise both today and tomorrow. By connecting the kinds of choices available for sea level rise to the legal frameworks that limit policy planning, this text should serve as an important framework by which meaningful adaptation planning can occur.

Chapter 2

Science behind Sea Level Rise

2.1 Introduction

Sea level rise is fundamentally a result of *chemical* and *physical* forces (Bryant 1997). While this statement is true, it does not provide an easy solution to the problems of sea level rise. Even so, without understanding the fundamental causes of sea level rise, we are less capable—and often wholly incapable—of making sound policy choices about how to act in response to sea level rise. The purpose of this chapter is to provide an overview of what science tells us about the potential causes and dynamics associated with sea level rise. A major goal in this chapter is to establish a foundation of the science behind sea level rise as a means of better understanding the connections between those scientific principles and the options we come up with as humans to respond to the impacts and effects of sea level rise. By understanding some of the scientific principles behind sea level rise, choices about policy directions can be made more meaningful and hopefully lead to better human decision making, the ultimate goal of this text.

The First Law of Thermodynamics, noting that energy is neither created nor destroyed, reminds us that no new water is being added to our Earth system; the same amount of water exists today as existed 50 years ago, suggesting that the phenomenon of sea level rise must be the result of causes beyond the addition of new water into the Earth system (Bertaianffy 1968). This chapter reviews the mechanisms behind sea level rise and discusses some of the geospatial (where things physically exist on Earth) considerations on how the physical characteristics of coastal areas impact sea level rise planning considerations.

Regarding causes of sea level rise, we have some understanding of the following potential contributors. Increasing temperatures are leading to a melting of water stored as ice. In addition, the increased heating of the oceans is causing sea water to expand (thermal expansion), further adding to sea level rise. Finally, human movement of water from largely stored underground sources (aquifers) may also be contributing to sea level rise as the water relocates to the oceans as part of the hydrologic cycle (Wada et al. 2012). The mix of warming and humans actively altering some of the normal sinks within the hydrologic cycle of water are providing feedback that is helping to drive sea level rise.

This chapter is divided into three main sections:

- Section 2.2, "Current Observations," aims to explore what we are seeing today in relation to sea level rise. The goal of this section is to help the reader understand some of the background science and observations that aid in our understanding of the dynamics behind sea level rise as an ongoing phenomenon. To aid in linking what we see today to a larger policy context, this section also begins to discuss the relevance of sea level rise in coastal management planning, particularly the question of *why* sea level rise deserves special consideration when thinking about coastal land use planning.
- Section 2.3, "Future Predictions," takes the information learned from Section 2.2 and asks two main questions in relation to that information. First, what is the potential for sea level rise in the future? Second, how does this impact land use planning within the coastal zone? The purpose of this section is to identify the importance of future predictions about sea level rise—essentially the importance of planning for the unknown by calculating the likelihood of future events—based on our current state of knowledge. The current state of knowledge about sea level rise serves as a foundation upon which future predictions about the extent of sea level rise are identified. Predictions about the future become the basis for planning efforts today and therefore are critical when making resource commitments; as such, the information relied upon when making such decisions is critically connected to our future actions. For example, the choice of whether or not to make infrastructure commitments in certain coastal areas is influenced by predictions about future sea level rise that are connected to our current state of knowledge. The choice to avoid infrastructure in a coastal area today will have an impact on the use of that area in the future. If the future prediction is unfounded, then the choice to avoid infrastructure can impact overall development and well-being today. Understanding the impact of future predictions on policy choices today is a key to understanding the risks involved in making planning decisions with less than perfect information. This relationship will be discussed in some detail within this section.

- Section 2.4, "Major Coastal Land Feature Considerations," explores the connections between the geography of different coastal areas and discusses the impact that local geographic conditions (spatial considerations) have on coastal planning. The simple calculus of rise over run (height over distance) is employed to add the dimension of geography as an important factor in coastal planning considerations. For example, a sea level rise of three-feet can have a substantial impact on an area where a three-foot rise in height can cover a very large distance of landmass, submerging that land in water. However, that very same three-foot rise in sea level can have little to no impact on an area where natural geography is mostly unaffected by the rise—for example, an area where the coastline is a bluff such as the White Cliffs of Dover in the south of England. Understanding basic geographic dimensions can have a significant impact on the overall planning process for dealing with sea level rise, and this section is devoted to connecting sea level rise planning to this important variable.

In summary, this first primary chapter is meant to provide the reader with an understanding of our current state of knowledge about sea level rise, and then relate that current state of knowledge to the process involved in making future plans about how we utilize our coastal regions. The connection between our current state of knowledge and the assumptions derived from that knowledge are emphasized as we consider the future predictions of sea level rise. Also, the geographic conditions of the coastal area in question help us place those predictions into context regarding how we think about impacts to coastal areas that geography will influence—either emphasizing or mitigating—those impacts. This preliminary understanding of the scientific principles behind sea level rise will then be used as a backdrop for the exploration of subsequent chapters in this text, particularly the policy options available to deal with sea level rise as well as the important legal considerations that influence potential policy directions.

2.2 Current Observations

Sea level rise is a phenomenon that is being observed all over the globe (IPCC 2007). Logical questions that follow from the current observation of sea level rise include: what is causing this observed rise in sea level; how much is sea level going to rise; what is the impact of sea level rise today and in the future; and finally, what, if anything, should we do about sea level rise? This section focuses on what we see today in terms of sea level rise. Subsequent sections get into the questions of future predictions and the impact of the particular geographic features of coastal areas. Before discussing the question of what is being currently observed, this section explores the question of why we care about sea level rise in relation to coastal planning in the first place—in particular, what role does sea level rise have in our discussion of coastal planning?

2.2.1 Why Is Sea Level Rise Something That Must Be Considered in Coastal Land Use Planning?

Coastal land use planning is a particular subset of land use planning generally. To understand coastal land use planning, including its purposes and goals, one must understand the purposes of land use planning generally, and then consider the particular conditions that occur in coastal areas that differentiate them from general land use planning considerations. In this review, one will find that coastal areas present a dynamic interaction between land and sea that is not normally encountered inland. This dynamic interaction has the ability to change boundaries between the land and sea interface, and also influence the impacts that forces of the sea have on human development at the nearshore.* Once a framework for understanding the unique circumstances presented in the coastal region is understood, the presentation of current observations relative to sea level rise will make more sense; this is particularly true as we move into the sections of this text that discuss policy options and legal considerations.

Traditional land use planning tends to focus on an orderly kind of planning surrounding human uses separated into different contexts based on a combination of preexisting uses and practical considerations (Randolph 2004). For example, planning for an already developed area may include considering how the land is currently being put to use and making determinations based on those existing uses; areas where industrial processes currently take place may continue to allow for industrial processes while residential uses may be restricted in this area to protect human welfare. Conversely, undeveloped areas allow for more planning options, and choices about development can be based on a broader set of considerations. Areas with superior views or located close to recreational opportunities may be best suited for residential development because of the demand for these kinds of natural attributes in residential living. Areas with less aesthetic attributes may be better suited for commercial or industrial purposes, depending on location, proximity to highways, and similar considerations.

One aspect of land use planning that is assumed to be consistent over time is the relative boundaries of land; this is one assumption that differs between traditional land use planning and coastal zone planning, particularly in areas where the coastal zone is subject to dynamic forces that can alter the boundaries between land and water. Unlike inland areas where boundaries are relatively stable, coastal zones contain a shifting boundary between the land–sea interface. A representation of this shifting boundary is identified in Figure 2.1.

As you can see from Figure 2.1, the boundaries between land and sea are based on the precise location where the land meets the water, which itself is constantly changing under natural forces of gravity, particularly the relative influence of

* The term "nearshore" here is being used to identify the area of dry land that is in close proximity to the waterline creating a demarcation between land and sea.

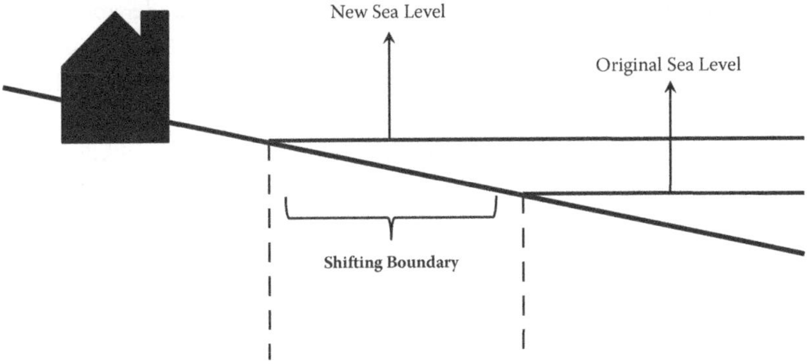

Figure 2.1 A representation of a shifting boundary between the land–sea interface.

the moon's gravitational influence over large bodies of water such as our oceans (Schlesinger 1997). Tidal shifts can alter the boundary between land and water, sometimes substantially in certain areas like the Bay of Fundy in Canada. We can refer to these cyclical changes in the land–sea interface along the coast from an average "high tide" to an average "low tide" as *tidal variation.**

Tidal variations are a normal part of the boundary changes between the land and sea. Tidal variations also provide some evidence of the special land use planning considerations that occur at the coast. The traditional method of defining the "set" line between the land and ocean is to establish a *baseline* that demarks the relative position between land and water for planning purposes.† Baselines are generally established by taking the average of the water line at a particular extent of tidal variation. For example, one can average out the mean high and low tides at a particular coastal location, and the baseline can be established as the average high or low tide mark at that particular location. Many legal standards internationally have adopted either the *mean high water mark* (meaning the average high tide water line of a given location) or the *mean low water mark* (the average low tide water line) as the basis for defining the official boundary between land and sea for land use planning purposes (Kalo et al. 2007).

The drawing of "official" baselines attempts to rationalize coastal land use planning by creating set boundaries so that the assumption of fixed boundaries adopted

* *Tidal variation* refers to the variation at a coastal point between the average high and low tides that are observed on a recurring basis. Influences such as moon phases and relative position of the Earth to celestial bodies can further impact the range of high to low tides, often referred to as *neap* and *spring* tides, respectively.

† It should be noted that the baseline is somewhat of an arbitrary construct, meaning that it aims to present a stable, nonchanging demarcation between the land and sea. However, the relative boundary between land and sea is constantly changing based on forces of gravity (and other factors) as described as the *tidal variation* above.

in general land use planning can be applied to the coast. It is understood that these boundaries are relative and fluctuations between the land and sea barrier occur regularly. However, an acknowledgment of natural fluctuations is bounded by the assumption that there is an upper limit between the highest of high tides and the lowest of low tides in a particular location. What is generally not acknowledged in the establishment of baselines is that other forces can irrevocably alter the historical baselines.

Coastal areas can be altered based on two primary considerations: (1) the background structure of the coast itself (discussed in a later section of this chapter), and (2) the environmental conditions impacting the coast. Land–sea boundaries can be impacted, sometimes irreversibly, by environmental conditions. Obvious examples include hurricanes or similar storms of significant energy intensity that shift sandy coastal areas to such a degree that the baselines between sea and land move substantially inland or seaward, depending on the impacts of the storm. Other examples are less discrete, taking place over longer periods of time. For example, erosion of certain types of coastal features (such as sandstone) can slowly remove land features at the coast, changing the land–sea boundary by impacting relative elevation of the nearshore. Also, the filling of submerged lands in one area—for example, the filling of coastal inlets and tributaries—can impact local shorelines by preventing the inflow of water into bays and estuaries, causing localized sea levels to rise in immediately adjacent areas (Philander 2008).

Environmental conditions can impact boundaries between the water and land. When environmental conditions change, they have the potential to change the underlying conditions upon which human decisions are made, and this is certainly true of the assumptions that go into establishing baseline boundaries between land and sea. Science plays a critical role in helping us understand how environmental conditions might impact our assumptions about land–sea boundaries, and this is particularly true when we consider the phenomenon of sea level rise as an environmental condition that is driving long-term changes in coastal boundary lines.

Sea level rise is being observed all over the Earth and it is having a disproportionate impact on coastal areas (IPCC 2007). As an environmental phenomenon, sea level rise has a variety of causes, many of which identify humans as a cause for the conditions that are leading to sea level rise. The causes and impacts of sea level rise are discussed in some detail in Section 2.2.2. Here we discuss the importance of sea level rise as something that must be considered in coastal land use planning. The main reason is that sea level rise brings with it a variety of factors that go beyond other environmental causes of changes to coastal boundaries. Importantly, the kinds of changes brought about through sea level rise at the coast can have the kind of impact and duration that make them permanent. As such, changes in coastal boundaries brought about by sea level rise require planning decisions that precede the changes; otherwise, coastal land use planning will devolve to a reactionary set of disaster response scenarios—not the best outcome when we think about the purpose behind planning for human interactions with land.*

As noted earlier, coastal boundary lines are ever changing, undulating landward and then seaward as the tides are influenced through gravitational forces of, primarily, the moon and sun. In addition, storms and other environmental factors (such as erosion) influence the boundary of land and sea, sometimes changing these boundaries in permanent ways.† However, current observed sea level rise is not the kind of environmental factor or natural background condition on which assumptions about coastal development generally rely. Instead, sea level rise represents a phenomenon that is unique to the current time frame in which human beings have flourished for the past 5,000 years and certainly unique to the current assumptions that define our law and regulatory regime related to land use planning, these legal assumptions being defined over the past several centuries to millennia in most cases. In this way, sea level rise represents a new problem that sometimes requires new ways of dealing with this problem at the coastal planning level of government interaction.

Sea level rise must be considered in coastal planning because it represents the greatest potential for change from the conditions we currently expect to find at the coast. From an impact standpoint, sea level rise presents new risks to coastal areas that go beyond the historical factors managers have dealt with in the past regarding coastal land use planning, development, and management. There are new *categories* of risk associated with sea level rise, including the continuous march of water inland over coastal areas now and for the foreseeable future. In addition, there are new *magnitudes* of risk as sea level rise represents, in part, a condition brought on by a warming planet.

The warming of the Earth including its oceans, increases the number and intensity of storms; coastal areas will experience storm impacts unlike those seen in the past. In addition, a warming planet presents powerful feedback in terms of energy dispersal that can impact coastal areas. For example, warm water melts ice sheets, thus increasing the amount of water flowing into our seas. In addition, warm water undergoes thermal expansion, which exacerbates sea level rise by causing the existing volume of ocean water to expand as it warms. The melting of sea ice exposes ground below the ice that is much darker than the ice itself; the reflective albedo effect of the ice is diminished and the darker ground underneath absorbs sunlight, thus reinforcing the warming effect on the Earth. Finally, a warming planet thaws permafrost areas of the Earth, releasing methane and other compounds into the

* Planning inherently suggests a proactive approach by humans to potential scenarios so the negative impacts of those scenarios can be mitigated (and hopefully often prevented) to the extent practicable. Planning that does not proactively respond to the potential impacts of sea level rise cannot rationally conclude to be meeting the essence of its "planning" goals.

† The term "permanent" here has a relative meaning, referring to a time period that may seem permanent relative to human beings, say, decades to centuries. However, the forces that alter coastline boundaries from a geological viewpoint (epochs, eras, eons) may seem impermanent in comparison to the shortened time frame of the human being's lens of perspective.

atmosphere reinforcing the greenhouse effect in our atmosphere and further adding to the warming trend.

Sea level rise as described above is very much an *indicator* of a warming planet. All the associated conditions relative to a warming planet have particular impacts on coastal areas. Increased storm frequency, intensity, and duration along coastal areas directly impact planning considerations within the context of how coastal areas are used and managed. Both ice melt and thermal expansion exacerbate sea level rise, and the continued approach of water inland impacts coastal management considerations. Finally, feedback that reinforces a warming planet makes it more likely that coastal area impacts mentioned above will continue in the mid- to long-term time frames of human existence. All these impacts influence coastal management in fundamental ways, primarily by ensuring a dynamic interaction between land and sea that is not bounded by what has been observed in the past. For these reasons alone, sea level rise is a phenomenon that requires special attention in coastal land use planning.

As stated above, sea level rise is the proximate cause of rethinking how we manage our coastal areas; the problem of rising seas incorporates direct challenges to the continued prosperity of coastal habitation. Indeed, there may be a breaking point at which coastal planning begins to rethink its policy options, including the planned retreat of developed coastal areas, as well as the purposeful non-development of coastal areas still untouched by human development. While the focus may be on sea level rise as a *symptom* of other activities that are reinforcing coastal policy choices, we must realize the *causes* of sea level rise are in many ways directly related to climate change. Because this text discusses coastal planning policies that accept the presumption that sea level rise is a given, it primarily focus on *adaptation* and *mitigation* strategies. However, it is important to note that other policy strategies exist, importantly the *prevention* of sea level rise. Any discussion of prevention options ultimately brings the focus back to climate change. This text does not focus on climate change policy options because the science indicates a high likelihood that sea level rise will continue for the foreseeable future (IPCC 2007). Still, readers should consider the impact that prevention strategies might have on the long-term impacts of sea level rise and at the very least connect the discussion of policy options contained in this text with options that take a wider preventative approach to the problem.

To understand the policy options and legal considerations chapters of this text, we must begin with an understanding of what science has to tell us about the current state of sea level rise today. Specifically, we need to know where sea level rise is currently occurring and what impacts it might be having on those areas where it is observed. After a short exploration of what is seen today, this chapter reviews future predictions related to sea level rise in a subsequent section as well as the role coastal land use features play in understanding the potential impacts of sea level rise. Armed with this descriptive information, we will then be capable of considering the policy options and legal implications connected to coastal management choices, making informed judgments about how to best manage coastal areas under an increasingly dynamic land–sea interface.

2.2.2 What Do We See Today?

As mentioned, there is growing evidence that sea level rise is occurring today. Observations over the past few decades have documented a sea that is continually encroaching on land in coastal regions across the globe. The Intergovernmental Panel on Climate Change offers the best overall summary of recent observations of sea level rise, linking those observations to concurrent increases in global ambient air temperatures. In summary, global average sea level has risen at an average rate of 1.8 millimeters per year between 1961 and 2003. The average rate of yearly sea level rise increase from 1993 to 2003 is about 3.1 millimeters per year, suggesting that the rate of sea level rise is increasing from past observations. Reinforcing the observed sea level rise are contributing factors that are also being observed. For example, the coverage of snow and ice observed over the globe has been decreasing in recent decades. In addition, the average temperature of the mid to upper areas of the Earth's oceans has increased based on data recordings since 1961, providing evidence that the oceans are absorbing additional heat, which in turn is causing the thermal expansion of sea water contributing further to sea level rise (IPCC 2007).

To put the current observations in perspective, historical evidence of sea level fluctuations can be used to compare the current observations of sea level rise and also to understand the potential extent of sea level rise. For example, this historical record indicates that global average sea level in the last interglacial period approximately 125,000 years ago (a warming trend in the Earth's history) was approximately 4 to 6 meters higher than currently observed sea levels; this translates into a 12- to 24-foot difference, mainly due to the melting of polar ice. Further evidence indicates the average ambient temperatures on the Earth were approximately 3 to 5 degrees Celsius warmer than average temperatures today. The direct implications of this evidence suggests that a warming of the average Earth temperature by 3 to 5 degrees Celsius can potentially have the impact of raising average sea levels across the globe approximately 12 to 24 feet (IPCC 2007).

One aspect of understanding sea level rise is to comprehend what is described above, the actual increases in sea level rise observed recently (1961 is very "recent" in geologic time but represents some of the earliest scientific attempts to accurately measure and quantify sea level rise) and the rates of sea level rise over that time. A more complete understanding of sea level rise comes from understanding the *causative factors* that come together to drive the currently observed phenomenon of sea level rise. For example, why is sea level rise occurring more frequently today? Is it possible that sea level rise has always been occurring on Earth and we just never began to directly observe it until recently? How certain can we be about sea level rise continuing to occur in the future even though we see it today; is it not just as possible that the rate of rise could begin to fall and even retreat? These questions are valid inquiries when we think about what sea level rise means in the context of policy decisions. It would seem unwise to engage in a policy direction that assumes sea level rise will continue only to see the rise level off and begin to retreat. What is needed is a deeper understanding of what factors are driving sea level rise and how those factors connect to the evidence (information) identified above to give policy makers valid information from which to make sound decisions. What follows now is an attempt to shed some light on the background causal factors of sea level rise in order to provide a framework for placing the phenomenon into a larger context to create a foundation for informed decision making.

There are two primary aspects to understanding what is being observed today: (1) the *extent* of sea level rise observed in different localities across the globe, and (2) the *impact* observed due to current sea level rise. The extent of sea level rise observed in a particular area is important because it provides evidence of the phenomenon itself, while also helping one understand the potential impacts locally. For example, an equal amount of sea level rise can have a disproportionate impact on certain coastal areas based on the geography of the particular region (discussed in greater detail in Section 2.4, "Major Coastal Land Feature Considerations"). Observing the extent of sea level rise in a particular location can help one identify the impact of sea level rise on that locality based on its unique makeup. These impacts can include readily observable phenomena on the surface such as inundation of coastal lands due to encroachment of rising seas. Impacts can also be hidden from immediate detection, such as the infiltration of sea water into underground aquifers along the coastline, or a rising water table due to such inundations of sea water in subsurface land (Wada et al. 2012). Understanding the extent of sea level rise can aid in better understanding the impacts, both actual and potential, it is having on a given coastal area.

2.2.2.1 Causes of Current Sea Level Rise

Earlier in this chapter, sea level rise was described as a *symptom* of other forces, chief among them was climate change. A more detailed discussion of the major causes of sea level rise beyond that quick summary is important to develop a foundational understanding of what is driving this phenomenon. To highlight the importance of this understanding, consider the role of coastal management from a temporal (time-based) standpoint. Coastal management decisions must be made today but with an eye toward the future; management decisions that only focus on what is observed today can create particular difficulties for future management and this is especially true when managing dynamic areas such as the coast. If coastal managers were only concerned with what is observed today, then decisions would be made that discount the impact of sea level rise, one example being the allowance of development near a land–sea border that contains coastal features sensitive to small increases in sea level rise.*

Understanding the connection between the causes of sea level rise and management decisions relative to the impacts of sea level rise, both today and tomorrow, are critical aspects of sound coastal land use planning. We now explore some of the causes of sea level rise in more detail in order to establish a firm understanding of the relationship between the science of explaining the phenomenon and the management decisions that flow from this understanding. This summary focuses on *energy flows* as a means of explaining the relationships associated with sea level rise. This process-oriented approach is chosen because it helps focus the discussion on the background environmental conditions without necessarily implicating any one particular condition as the sole cause of observed sea level rise. In addition, understanding the issue from an energy flow perspective should provide a sense of the systems dynamics that are involved in regulating the Earth's processes. Such a systems approach is an excellent overall foundation from which policy frameworks can be developed in the coastal zone.

The primary cause of sea level rise is a change in the regularly observed state of the *Earth system*, often referred to as an *equilibrium state* (Bertaianffy 1968). There are two important concepts identified here: (1) the first concept suggests that the Earth operates as a *system*, and (2) the second concept suggests the Earth system is in a state of existence that is at *equilibrium*, meaning conditions are relatively stable. These concepts should be discussed in detail in order to gain a better understanding of their meaning in context; we will look at each concept individually to ensure there is a solid understanding of these principles. A more in-depth discussion of how these principles apply to our understanding of the causes of sea level rise will follow.

The concept of the Earth as a *system* is well accepted in scientific fields, finding its most specific theoretical underpinning in the field of ecology (Molles 2008). Applications of the Earth as a system are used to understand interactions in other earth sciences such as geology and interdisciplinary process-oriented subjects such as biogeochemistry (Schlesinger 1997). The Earth acting as a system is meant to suggest a set of processes that occur within a given space (McGuire 2012). A simple representation of a system is shown in Figure 2.2.

In the case of the Earth as a representative system, the global dimensions of the Earth itself establish the boundaries of the system. The processes (interactions) that occur within the Earth influence how it operates and thus what it looks like. For example, within our Earth, we have different areas upon which energy interactions

* An example would be a coastal area that is low lying and has a very slight elevation over a long distance: very little *rise* over a relatively long *run*. A 0.5-foot increase in elevation for every 100 feet of distance from the shoreline would result in a 300-foot inundation of sea water for a 1.5-foot increase in sea level. Thus, allowing for development within 300 feet of the shoreline where sea level is likely to rise 2 feet within the next 50 years would be a poor policy decision looking ahead 50 years even where it may seem like a good decision today.

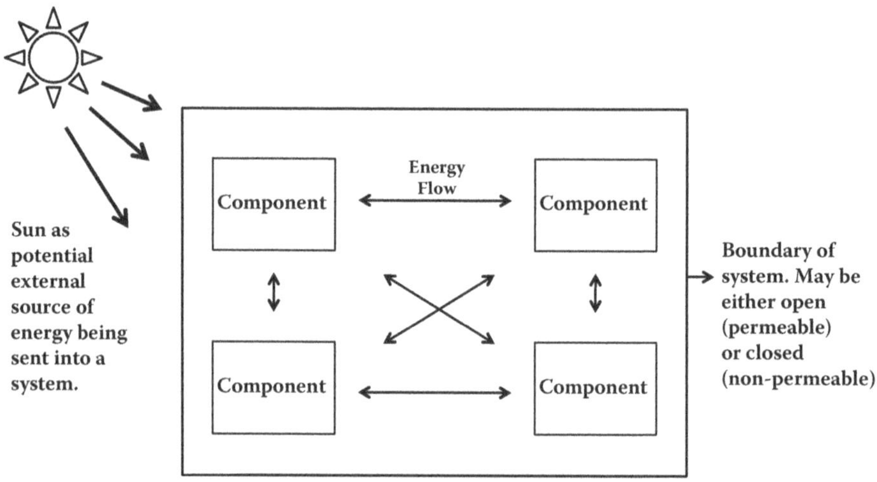

Figure 2.2 A representation of a system.

occur. There is the biosphere (all living things), the hydrosphere (water), the atmosphere (air), and lithosphere (land) (Molles 2008). There are interactions between these different spheres that influence one another. The amount of water taken up into the atmosphere influences how much rain is available to fall onto the land, which then impacts how the land looks—for example, the difference between desert land (little to no rain) and rainforest (lots of rain). The different spheres, or components, of the Earth system interact with one another and thus influence each other. Collectively, these influences impact how the Earth looks and operates. This is the concept of the Earth as a system in a nutshell.*

The other major concept discussed is that the Earth is at a state of *equilibrium*. This essentially means that the components of the Earth system have been interacting with each other for a sufficiently long period of time so that the Earth itself has become well mixed. Being well mixed means the concentrations or abundances of things found on the Earth are essentially constant. It is like a well-mixed shake that is full of different ingredients, some liquid and some solid. When the shake is well mixed, there are approximately equal amounts of the different ingredients throughout the shake; if you sample the ratio of ingredients at different areas of the shake—say, top versus bottom—they should be about the same. However, if you allow the shake to sit without mixing for a long period of time, you might notice that the ingredients begin to separate, with the more dense solid materials sinking to the bottom and the less dense liquid materials moving toward the surface. The Earth system is generally presumed to be in a state of equilibrium because

* James Lovelock developed a theory about the Earth as a living system. He proposed a hypothesis that said the Earth was self-regulating as an entity based on the interactions between the components (spheres) of the Earth system.

the components of the system have been constantly interacting with one another for a long period of time, so what we observe today is generally what we expect to observe tomorrow: this is the basic concept of equilibrium.

A change in the interactions between Earth system components can bring about a change in the equilibrium state of the Earth system itself. The *cause* of these changes is almost always a shift in energy distributions within the system (IPCC 2007). The causes of sea level rise can be discussed in the context of Earth system dynamics, particularly by looking for shifts in energy distributions throughout the system. The clearest way to explain energy shifts within the Earth system itself is to look for changes in the major components of the Earth system; by identifying shifts in energy distributions between system components—say shifts in the hydrosphere and atmosphere—the cause of sea level rise becomes apparent.

The first step in this analysis is to focus on one component of the system and identify what is generally observed and then to take note of changes in what is normally seen. In the case of the hydrosphere, we generally see a relative distribution of the water on Earth in various phase states: liquid, solid, and gas (Molles 2008). These several states exist within the range of temperatures observed on, above, and within the Earth's surface. Throughout recent human civilization, we have generally observed large bodies of water in liquid form in the oceans, lakes, rivers, streams, and underground aquifers. As we move toward the poles of the Earth or increase our elevation, we tend to see water in its solid state. Finally, we sometimes observe heated water from below the Earth's surface being released through geysers and hot springs, some of which is released in the form of gaseous steam vapor. There are small variations to these normal observations. For example, there are seasonal fluctuations in temperate regions of the Earth at mid and higher latitudes from the equator. In these temperate regions, water fluctuates in states from liquid to solid depending on the season (the relative distance from the sun), with winter allowing for water to exist in a solid state of snow and ice, and summer allowing water to remain in a liquid state. Taken collectively, this is the pattern we have observed in our Earth system during approximately the past 10,000 years of human civilization.

The observed pattern of the hydrosphere summarized in the previous paragraph is entirely dependent on the flow of energy throughout components of the Earth system. The major source of external energy driving the interactions between components of the Earth system is the sun.* The heat portion of the electromagnetic energy that comes from the sun "drives" many of the physical forces we observe on Earth, including ocean currents, wind patterns, and other physical processes. The distribution, intensity, and residence time of solar heat throughout the planet is ultimately responsible for the current patterns we observe and have come to rely on

* The Earth also has a substantial internal energy source, as the internal mass of the Earth is made up primarily of molten magma. This energy drives movements of the Earth's crust (plate tectonics) and is also responsible for "hotspot" island building, volcano eruptions, geysers, hot springs, and other related phenomena.

throughout our development. For example, the *trade winds* are a prevailing west-to-east wind pattern that is almost always found in the mid latitudes of the northern hemisphere. They are called "trade winds" because they were used as a kind of highway system for sailing ships engaging in trade along the Atlantic Ocean between Europe and the Americas. The prevailing winds—a critical component to sailing as wind provides the power source for movement—could always be relied upon to move in the west-to-east direction, meaning that sailors could depend on this wind direction when planning their routes. The distribution of heat energy provided by the sun creates temperature differentials on the Earth's surface, and these temperature differentials create particular wind patterns at different latitudes.* The result has been a relatively stable pattern upon which human civilization has identified and depended for its development.

Changes in observed heat distribution patterns can impact components of the Earth system such as the hydrosphere. Lower-density hot air rises at the equator and then cools as it moves up into higher elevations of the atmosphere. Wind patterns push the denser cooler air to higher latitudes where it begins to sink. Some of the air is distributed back toward the equator, while other portions are heated and rise to yet higher latitudes. Collectively, these patterns help to move currents in the ocean as warm air masses transfer heat to the oceans and that heat then drives currents through a process of cooling and heating that alters water densities and thereby creates movement as a by-product (Schlesinger 1997). This process is not unlike how our bodies regulate heat by using our system of blood vessels to move warmed liquid blood to the surface of our skin where it can be cooled by the air outside our bodies. The Earth engages in a set of responses to the heat energy that comes into the planet, and the sum of those responses over a great length of time has helped to create the current state of equilibrium we now observe. But what happens when the amount of heat trapped in the Earth system increases over time? What impact might that extra heat have on a component of the system such as the hydrosphere? By answering this question we come closer to understanding one of the major causes of sea level rise.

Recall that the Earth has been described as a system with a set of inflows, interactions, and outflows as the major processes that define the system. Heat energy from the sun has just been described as a major inflow into the Earth system causing interactions through the driving of processes as described above. What has not been described in detail is the outflow portion of this process, meaning what happens to the heat energy that comes into the planet from the sun? If all of the heat

* There are other factors involved in this process, most of them relying on physics. The general oval shape of the Earth, along with its slight axial tilt, means heat energy is not absorbed on the Earth equally. In general, the equator gets a constant rate of heat energy, while the amount of heat absorbed changes seasonally as one moves to higher latitudes from the equator. In the summer, more heat is absorbed by higher latitudes because the axial tilt allows the Earth to be closer to the Sun's rays. In the winter, less heat is absorbed because the higher latitude mass of the Earth is further away from the sun.

stayed on Earth, and new heat energy continued to flow into the Earth every day from the sun, then the Earth would have to find ways to deal with this heat. Some of the heat may be stored in different parts of the Earth; liquid water has an amazing capacity to store heat without undergoing a phase change and the surface of the Earth is made up of mostly water. However, without going into too much detail, we can assume there is a point where the ever-increasing ingestion of heat by the Earth will impact the system; call this point an upper limit on the ability of the Earth to continually allow for inflows of new heat without any outflows of that heat. At some point, the increased flow of heat into the system should cause changes to the interactions within the system. If changes are observed, then those observations may be evidence that the system itself is not functioning in its preexisting equilibrium state. Knowing what is happening to the outflow of heat into the system is one way to help determine the cause of observed changes within the system.

Not all heat energy that flows into the Earth system remains within the Earth. Some of the heat energy is immediately reflected back into space (outside the Earth system) by clouds, ice, snow, and other light-colored objects that tend to reflect the wavelength of electromagnetic energy that contains heat. This reflective capacity of light-colored objects to heat energy is often referred to as the *albedo effect* (Schlesinger 1997). Some heat is taken up by interactions with the ocean where water absorbs and dissipates the heat over massive volumes. Other heat is dissipated into the atmosphere after interacting with the Earth, ultimately leaving the upper levels of the atmosphere and exiting to outer space. One inhibitor of heat escaping into the atmosphere includes the chemicals that are found in the atmosphere. Some combinations of molecules create compounds that help form a barrier to heat energy leaving the atmosphere. One element that can help form these heat-trapping compounds is carbon. Examples of heat-trapping compounds include carbon dioxide (CO_2) and methane (CH_4). When these compounds are found in higher concentrations in the atmosphere, more heat tends to be trapped in the Earth system; the compounds allow new heat to enter the system but they prevent existing heat from leaving the system. This phenomenon is commonly referred to as the *greenhouse effect* (Schlesinger 1997).

The greenhouse effect can cause the Earth system to carry more heat than is historically found within the Earth since human civilization began. The effects of more heat within the Earth system can be broad and multifaceted. With respect to the hydrosphere, more heat will generally cause higher ambient temperatures. Higher temperatures can cause the relative equilibrium state of the hydrosphere to shift, for example by melting greater amounts of water stored as ice and having that water run into the oceans. The ultimate cause of this "heat trapping" on the Earth is increased carbon in the atmosphere. The increased carbon in the atmosphere is directly caused by perturbations of the carbon cycle, another part of the Earth system.

The carbon cycle represents how carbon molecules are utilized within the Earth system. Relying on the laws of thermodynamics—particularly the law that notes energy is neither created nor destroyed—the carbon cycle suggests that the amount

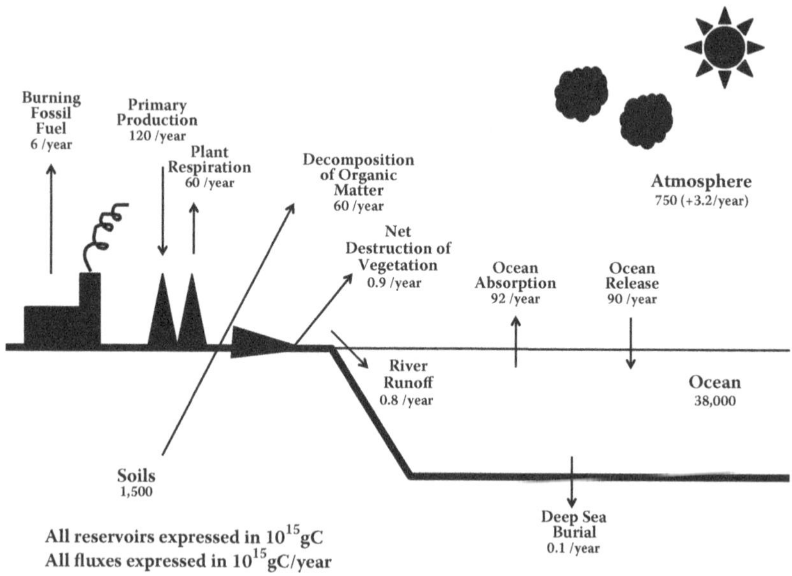

Figure 2.3 A representation of the carbon cycle. Data from William H. Schlesinger, *Biochemistry: An Analysis of Global Change*, 2nd ed. (San Diego, CA: Academic Press (1997).

of carbon on the Earth is essentially constant (Schlesinger 1997). Moreover, in the well-mixed Earth system, carbon concentrations are at equilibrium, meaning the amount of carbon within any component of the system is relatively stable and at a certain concentration that will be the same even when measured at different times. This idea of well-mixed concentrations is readily observed in the atmosphere with oxygen; the concentration of oxygen in the atmosphere is around 21 percent and it remains relatively constant over time even though the oxygen itself is constantly being removed and added to the atmosphere (Schlesinger 1997).

The concentration of carbon in the atmosphere has been increasing in recent decades, with the main cause being the movement of carbon to the atmosphere through the removal and burning of buried carbon from the lithosphere. To understand how this is occurring, a representation of the carbon cycle is shown in Figure 2.3.*

Carbon cycles between different components of the Earth system, spending time in certain components of the system and then moving to different components of the system. Following a single atom of carbon can provide a telling story of how carbon cycles throughout the entire Earth system. Imagine carbon that is stored in the body of a plant. The plant is consumed and, as a result, the carbon is sent

* For a detailed summary of carbon cycling generally, see McGuire, Chad. 2012. *Environmental Decision-Making in Context: A Toolbox*. Boca Raton, FL: CRC Press. (p. 51–57).

into the atmosphere through a process of respiration. Once in the atmosphere, the carbon atom links up with oxygen to form carbon dioxide. The carbon ultimately gets taken up by a plant near the surface during photosynthesis and is transformed into a simple carbohydrate, becoming part of another plant body. The plant is eaten again and the carbon atom becomes part of the consumer's muscular system. The animal (say, a dinosaur) dies in a low oxygen environment and its body slowly decays over millions of years, ultimately becoming oil trapped underground. The carbon atom is now oil that is stored under the ground (lithosphere) where it will spend a long time buried and therefore not interacting with the other components of the Earth system—that is, until humans drill through the ground for the oil, refine and burn the oil, sending the carbon into the atmosphere.

The story of the carbon atom cycling throughout the Earth system helps identify an important contributor to sea level rise. To understand carbon's contribution, a summary knowledge of system dynamics and equilibrium theory is necessary. With this summary knowledge, the following can be surmised: The Earth's carbon cycle is shifting out of balance with its historical equilibrium state through actions that include the removal of carbon stored in one component of the system (lithosphere) and the addition of that carbon to another component of the system (atmosphere). This added carbon, in turn, strengthens the greenhouse effect of the upper atmosphere when that carbon combines with oxygen and hydrogen to form "greenhouse gases" (carbon dioxide and methane as examples). The greenhouse effect is preventing the sun's heat energy from escaping the atmosphere and causing the ambient temperatures at the surface of the Earth to rise. This rise in surface temperature is stimulating a phase change through the melting of water in areas where it has traditionally been stored as ice. The melting is causing more liquid water to enter our oceans, thus causing ocean levels to rise. This process is similar to how leaving the faucet on in a plugged bathtub will cause the water level in that bathtub to rise.

So there we have it: The relationship between energy flows and the Earth system is responsible for some of the sea level rise we are currently observing, and most likely the vast majority of sea level rise currently observed (IPCC 2007). The main culprit for this change is the movement of carbon stored below the Earth's surface to the atmosphere outside of the normal background rates where this might otherwise occur in the Earth system. To understand the difference, consider that carbon cycling throughout the Earth system has been well mixed over time; the set amount of carbon is at a state of equilibrium throughout the Earth as represented in Figure 2.3. Under background conditions, carbon stored under the Earth's surface might remain there for thousands to millions of years before it is "released" in some manner back into another component of the Earth system (hydrosphere, atmosphere, etc.). When the carbon is actively mined from long-term burial and moved to another part of the system, the buried carbon gets added to the existing carbon in that component. In the case of the atmosphere, the concentration of carbon exceeds background rates. Because carbon can combine with other elements in the atmosphere (particularly an oxidizing atmosphere, one full of highly reactive oxygen) to form gases capable of trapping heat, the Earth system responds to this new stimuli by holding onto more heat energy than it normally would with less carbon in the atmosphere.* If the amount of carbon in the atmosphere remains higher than background rates over time, then one might hypothesize that the Earth would continue to experience an increase in ambient surface temperatures to some point of system saturation.

There is evidence of how far a greenhouse effect can go in impacting a planet system. The planet Venus shows signs of what is referred to as a "runaway" greenhouse effect. Temperatures on the surface of Venus far exceed those of Earth, and Venus has a much denser atmosphere. The dense atmosphere acts as a super-insulator of heat, creating a very strong greenhouse effect where heat is trapped and stored within its system for long periods of time (Margulis and Sagan 1995). Some scientists argue that Venus is a good example of how feedback mechanisms can help reinforce and even accelerate a condition within a system, sometimes drawing the system outside its background equilibrium state and beyond a threshold where a new equilibrium state is reached. A schematic representation of movement toward new equilibrium states is shown in Figure 2.4.

The argument for the Earth is that, like Venus, the addition of heat-trapping molecules in the atmosphere can reinforce conditions that make the planet hotter, for example by removing ice, snow, and other light-colored objects that reflect heat. This, in turn, can cause greater evaporation of water, which leads to higher concentrations of water vapor in the atmosphere, which then leads to more trapping of heat in the atmosphere,† reinforcing the warming of the planet. If this cycle continues unchanged, then the Earth system can pass a threshold, as shown in Figure 2.4. When a threshold is passed, the system can enter into a new equilibrium state. The result is that the Earth looks very different from what we are used to. This is what some scientists argue happened to Venus: The planet underwent an equilibrium shift due to feedback that reinforced the trapping of

---

* Assuming the stability of all other factors. For example, this statement is true if the amount of heat energy from the sun entering the Earth system remains constant. If the amount of heat energy were to increase from the sun, say through a burst of solar flares, this would also impact the amount of heat observed on the Earth in relation to background conditions. Similarly, if the amount of heat energy from the Sun were to decrease in relation to historical rates of heat energy supplied, then this would also impact the effect of heat-trapping compounds in the atmosphere; less heat input into the system would counteract some of the effects of additional heat-trapping compounds in the atmosphere, thus affecting the "net" amount of heat being stored in the Earth system during any particular period of time. Understanding the feedback dynamics of the system helps us understand causal relationships between variables that can impact the condition of the system; for example, less heat coming in from the sun can mitigate the effect of additional heat-trapping compounds found in the atmosphere.

† Water vapor is a greenhouse gas because it has a high capacity to hold heat. The more water vapor in the atmosphere, the greater the capacity of the atmosphere to hold heat energy. Some of the heat generated through water vapor accumulation in the atmosphere can be offset by greater cloud cover, which can reflect heat before it enters lower parts of the atmosphere.

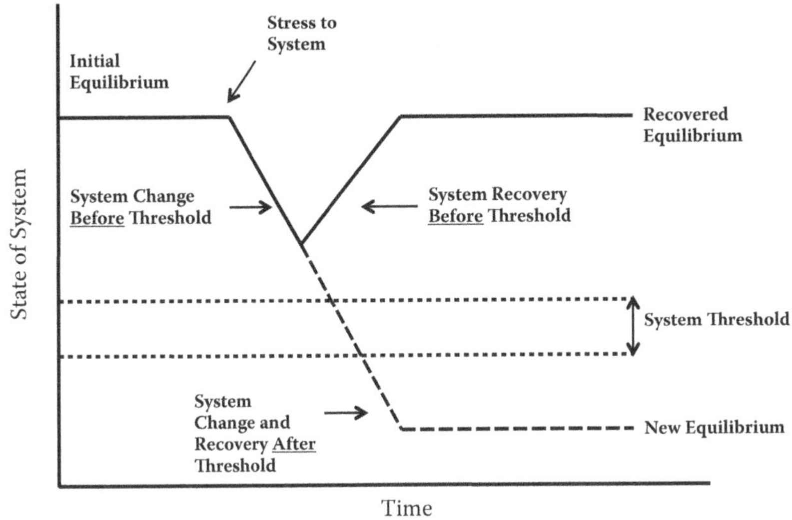

Figure 2.4 A representation of a system moving to a new equilibrium state.

heat within the planet's system and it is now a very hot and inhospitable place for life—at least life as we know it.

Let us now take a moment to summarize the major points made thus far in this section. Recall that the goal here is to get a sense of what we see today in terms of sea level rise. To understand the current observations, some explanation of the science behind sea level rise is warranted. The previous discussion placed the science behind sea level rise into the context of systems dynamics and equilibrium theory, suggesting the Earth acts as a system with inputs, interactions, and outflows and has been doing so for a sufficiently long period of time that the entire Earth system is well-mixed, meaning that the concentrations observed and rates of change within the components of the system are relatively constant. Humans are impacting the normal distribution of parts of the system, particularly carbon. By removing stored carbon from the ground and placing it into the atmosphere, humans are increasing the capacity of the Earth system to hold heat energy from the Sun. The additional heat is raising the ambient temperature of the Earth's surface and a side effect is the melting of ice. The melting of ice is one of the major contributions to current observations of sea level rise. The science behind a warming trend observed on the Earth helps to explain the concurrent observation of rising seas. Knowing the science can be helpful not only in understanding the cause of sea level rise, but it can also aid in the development of policy prescriptions to manage and mitigate the impacts of sea level rise.

Water added to the oceans through melting ice and snow is not the sole cause of observed sea level rise. Other forces are contributing to current observations of rising seas. For example, *thermal expansion*, the slow increase in surface and mid-water temperatures in the ocean, is hypothesized to be adding to the rising level of oceans (IPCC 2007). There is also some recent evidence that supports the proposition that human removal of fresh water from underground aquifers is contributing to sea level rise (Wada et al. 2012). This theory, supported by calculations of groundwater removal globally, suggests that this particular source of water is being removed from its normal background rates of underground storage by human effort. This water is being moved from one part of the hydrologic cycle (underground storage) and ultimately being deposited in another (the oceans). This process is occurring in a way that is influencing the equilibrium state established in the hydrologic cycle.

There are, undoubtedly, multiple feedback mechanisms impacting the current observed phenomenon of sea level rise. However, there is strong certainty that climate change is a substantial contributing factor. As mentioned earlier in this section, current observations of sea level rise are broken into two main categories: (1) the *extent* of sea level rise observed, and (2) the *impact* of the current observations. A summary of the current extent of sea level rise was detailed at the beginning of this section, followed by a discussion linking the scientific principles to help in understanding *why* sea level rise is occurring. The goal now is to look at the *impact* that current observations of sea level rise are having on our coastal areas. Current observations of impact will be extended in Section 2.3, where future predictions of sea level rise are discussed in greater depth, including the impacts that future predictions might have on land use planning within the coastal zone.

2.2.2.2 Impacts of Current Sea Level Rise

Current impacts of sea level rise depend in large part on the land features encountered along a particular coastal region; note that coastal land feature considerations are discussed in greater depth in Section 2.3. So as impacts are discussed generally here, qualifications will be made on the *potential* for these impacts to be realized in relation to different features encountered at the coast. Qualifications aside, current impacts on coastal areas can be divided into three main considerations: (1) *physical* impacts that take into account the physical features of the coast and the physical processes that both drive sea level rise and are impacted by sea level rise; (2) *ecosystem* impacts that consider the effect of sea level rise on vulnerable coastal species and associated habitat; and (3) *societal* impacts that consider the effect that sea level rise has on the human-built environment.

Physical impacts to the coast vary, depending on several factors, including elevation and proximity to storm centers (Brooks, Adger, and Kelly 2005). In general, lower-lying coastal regions with a gradual slope are more at risk than higher coastal elevations and areas with a more dramatic slope. In addition, the proximity of the coastal area to epicenters of strong storm activity (e.g., hurricanes, tsunamis, earthquakes, etc.) is a factor for increased physical impacts as sea level rises. Also, the depth of the water impacts the ability of ocean-borne storm surges to reach closer to the shoreline before releasing their energy; greater ocean depths brought on by sea level rise can increase a storm's proximity to coastal land before releasing the brunt of its force. Warming ambient surface temperatures increase the temperature of surface waters, which in turn increase the intensity of storms such as ocean-borne hurricanes. Warmer surface waters moved inland through sea level rise increase the probability and reach of coastal storm impact (Schlesinger 1997).

The expected physical changes associated with coastal areas include the flooding of low-lying areas. Flooding, or inundation, generally occurs in sheltered, low-energy areas where physical processes such as sediment accumulation are minimal (U.S. Global Change Research Program 2009). An example of protection against flooding in a low-lying area includes the levees built in New Orleans, Louisiana. The levees created an artificial environment where the water's edge buttressed up against human development. When Hurricane Katrina struck in 2005, the sea was pushed inland—albeit by a storm surge—unabated by natural sediment buildup or wetlands. The result was the immediate inundation (submergence) of large portions of the city nearest to the levees. This example helps show the kind of impact physical processes can have through sea level rise, including what can currently be observed across the globe. The extent of inundation is primarily based on the relative slope of the land where protected coastal areas with minimal slopes have the greatest potential to be impacted by sea level rise. Specific principles associated with the geography of a coastal area are discussed later in this chapter.

Beyond inundation, the movement of the sea inland causes land loss through erosion of emerging land–sea boundaries. As the sea moves landward, it creates shifting coastal boundaries. Some of the upland connected to the ocean can be more susceptible to erosion, depending on the makeup of the soil and the dynamics associated with the new land–sea boundary. A strong surging tide can also impact the shape of the nearshore landmass, creating shifting land that is relatively unstable. In addition, the rate of sea level rise in a given coastal area can impact existing wetlands and the development of new wetlands or intertidal areas, depending on local conditions (Brooks, Adger, and Kelly 2005). Approaching sea water can also mix with existing nearshore underground aquifers, impacting freshwater resources. In addition, the physical impacts of rising seas can also impact the ecological and societal concerns in coastal areas.

This quick summary of physical impacts suggests a few insights relative to current observations of sea level rise. First, sea level rise is a condition that occurs from physical processes associated with heat energy exchange; when sea level rise is observed, the other associated physical processes must be considered, including the physical impacts those associated processes can have on coastal areas. Second, physical processes tend to have the greatest impact on coastal areas that are relatively sheltered and low lying with a slight gradient. Third, the physical impacts to coastal areas that maintain human development and have employed armoring techniques, such as building levees to hold back the sea, can be particularly vulnerable to sea level rise and its attendant risks. Many of the physical impacts of current sea level rise are already being felt along costal regions. As mentioned earlier in this chapter, global sea level has been observed to be rising recently in the late nineteenth and early twentieth centuries (Lambeck et al. 2004; IPCC 2007; Gehrels et al. 2008). Recent analysis has shown that global averages of sea level rise have been approximately 7.5 inches during the twentieth century (Jevrejeva et al. 2008) and the rate of sea level rise has been increasing over the past 15 years (IPCC 2007). Finally, there is strong evidence that the rate of sea level rise will exceed that of the twentieth century in the twenty-first century (Jevrejeva et al. 2008). Areas of the coast that exhibit combinations of at-risk characteristics identified above and are found in dynamic locations will be prime targets for continued and sustained physical impacts of sea level rise for the foreseeable future. As such, these are likely key areas for policy consideration.

Ecosystem impacts consider the effect of sea level rise on vulnerable coastal species and associated habitat. A *coastal zone* often contains a variety of characteristics that are consistent between coastal features across the globe. These features include a water line that distinguishes between land and sea, an intertidal zone that represents the fluctuation of the sea–land interface based on tidal range, an area above the influence of the tide (sometimes a sandy beach), and a vegetation line that represents the landward extent of the coastal area. A representation of a coastal zone is shown in Figure 2.5.

Many coastal zones contain *wetland* areas: nearshore zones that are either always partially wet or intermittently wet, and usually subject to tidal influences. Wetlands can contain unique plant species that help form important habitat for a variety of marine and terrestrial animals; coastal wetlands are often considered ecologically important zones because of the richness and diversity of species present

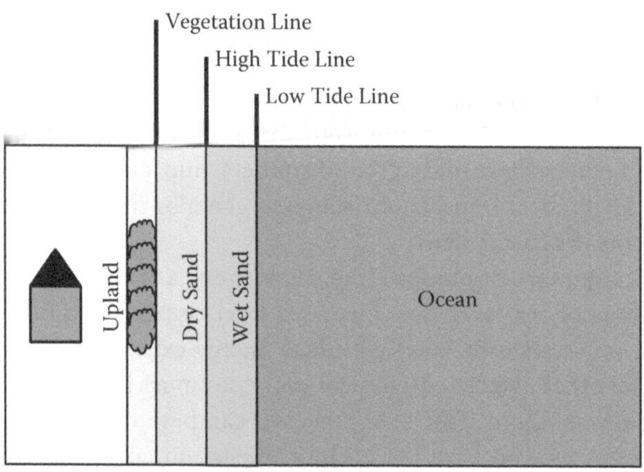

Figure 2.5 A representation of coastal zone features.

(Costanza et al. 1997). As sea level rises and inundates inland areas, the normal ecological footprint of wetlands changes. Areas that were partially submerged become completely submerged. Plants that rely on close proximity to water but are not adapted to full submergence die off as the water overtakes them. In sum, sea level rise changes the balance of wetland ecosystems, and this change in balance can disrupt the functions and processes of wetland ecosystems.

The extent of the damage to ecosystem functions from sea level rise depends on a variety of factors, one being whether or not wetlands are capable of migrating inland naturally. If sea level rise occurs at a relatively slow pace on open land that can accommodate wetland migration, then it is likely that many of the characteristics of the coastline will be retained as water moves inland; wetlands will slowly move inland as the steady rise in sea water landward provides the conditions that are favorable for wetland species (Bryant 1997; Riggs and Ames 2003). However, if the land is armored against the rising sea—for example, through the building of a concrete seawall to hold back the water—then there is little to no capacity for a traditional coastal zone to develop, including wetlands, as generally described in Figure 2.5. The impact to the local ecosystem due to wetland loss through sea level rise can be substantial as the mosaic and richness of habitat provided through wetlands are no longer available to support a diversity of life; sea water abuts directly a human-made structure, creating a sharp contrast between land and sea, salinity, and terrestrial-versus-aquatic habitats.

Current sea level rise is showing an already dramatic impact on ecosystems. For example, in low-lying regions of the United States, ocean storm surges and particularly high spring tides are increasingly causing flooding of nearshore lands. These events are causing the loss of wetlands and the conversion of nearshore forest and developed lands (farmlands, residential properties) to wetlands (Riggs and Ames 2007). Freshwater areas (lakes, ponds, aquifers) that are found near the land–sea boundary are also increasingly becoming flooded with salt water from sea level rise, which is changing the salinity of the freshwater bodies leading to significant impacts on those localized ecosystems (U.S. Global Change Research Program 2009).

The impacts on these ecosystems and the services they provide can be substantial in direct and indirect economic terms. As mentioned earlier, coastal wetland resources provide valuable nursery habitat for both commercial and recreational fisheries (Costanza et al. 1997). In addition, coastal wetlands provide important buffers from ocean-derived storms, mitigating the intensity and impact of storm surges on coastal areas (Bryant 1997). There are other valuable provisioning, regulating, and aesthetic services provided by coastal wetlands (Millennium Ecosystem Assessment 2003); the value of these services has been estimated to be in the billions to trillions of dollars (Costanza et al. 1997). The fact is that current impacts of sea level rise on the ecological values of near-coastal areas are substantial, and the impacts will likely increase with continued sea level rise. Importantly, choices about development in undeveloped coastal areas today can have a substantial impact on the extent of ecological impacts in the future. As mentioned earlier, the decision to armor against the rising sea by building walls prevents the natural continuation and development of coastal features inland, most notably coastal wetlands. Preventing wetlands from migrating as the sea rises will only intensify the loss of ecological values, and this should be clearly identified in any policy direction taken in response to sea level rise; coastal management needs to incorporate the important ecological values at stake in the coastal zone, particularly values that are not directly associated with human development at the coast.

Societal impacts observed from current sea level rise range depending on a variety of factors, but two factors correlate highly with overall societal impact: (1) geographic vulnerability to sea level rise and (2) population density around coastal regions (Adger, Vincent, and Vincent 2005). Broadly speaking, low-lying island nations that are particularly vulnerable to sea level rise are experiencing a disproportionate impact on social and cultural identity. This is primarily due to the fact that "vulnerability" related to sea level rise is often determined by two factors: (1) exposure to a hazard and (2) the capacity to adapt to the hazard. Beyond island nations, low-lying coastal communities with available upland retreat areas can be just as vulnerable to sea level rise as island areas where retreat inland is not possible, and sometimes these regions can be more vulnerable than island nations. The issue is really one of considering both the environmental factors (topography, etc.) that present objective conditions of risk, and also considering the human-based factors of risk that center on the capacity of social institutions to prepare and adapt to rising seas.

Vulnerability suggests a mix of actual environmental risk (for example, the low-lying nature of an area providing the non-human controlled portion of risk) and human decisions related to those risks. How humans decide to plan and respond to risk—including assessing the capacity to plan and respond to risks—is an important human-defined factor in determining the vulnerability of a region. Human-based factors impacting this kind of risk assessment of vulnerability include awareness of the hazard, intensity and sophistication of development along coastal regions, and the kinds of public policy institutions established to deal with the risk (Brooks et al. 2005). As Anthony Oliver-Smith points out, "[v]ulnerability…explicitly links environmental issues, such as hazards, with the structure and organization of society, and the rights associated with membership" (Oliver-Smith 2004, p. 15).

Vulnerability thus links actual environmental hazard to human-based institutions and the manner in which those human institutions have provided for the needs of its society; where institutions are weak or allow for inequitable distribution across its society, the vulnerability of those citizens will generally be higher than in a society with stronger institutions and more equitable distribution of rights and benefits among citizens. This is true even where the environmental factors are similar between nations—for example, where two nations share similar exposure to sea level rise and similar population densities near coastal areas but have different socio-political structures in place to deal with the threat of sea level rise. The nations with stronger socio-political structures and institutions will generally have less vulnerability toward sea level rise than the nation with weaker institutions (Nichols and Hoozermans 2005).

In terms of what is observed today, low-lying island nations with high population concentrations along the coast are generally experiencing a disproportionate exposure to the hazard of sea level rise; and by virtue of being a relatively small body of land surrounded by water, these island nations also tend to have a lower capacity to directly adapt to the hazard of sea level rise in their current locations. Small Island Developing States (SIDS) include island nations such as Tuvalu, Kiribati, and the Carteret Islands. These island nations are currently experiencing sea level rise and associated impacts of the kind that is threatening the continued existence of their physical landmass (Barnett and Campbell 2010). These island nations, as well as many other nations defined as SIDS, contain characteristics that add to the vulnerability from sea level rise: They have a *small size* (limited natural resource base, high competition between land use, intensity of land use, immediacy of interdependence in human–environment systems, spatial concentrations of productive assets); they are *insular and remote* (including geopolitical weakness); they have similar *environmental factors* (small exposed interiors with large coastal zones); they contain *demographic factors* that increase vulnerability to sea level rise (limited human resource base, small population, population concentrated within coastal zone; they have single economy characteristics such as tourism); and the *economic factors* of these countries tend to increase overall vulnerability (small economies, small domestic market, dependence on natural resources, highly specialized production) (Pelling and Uitto 2001).

Island nations can experience disproportionate risks toward sea level rise compared to other nations with coastal exposure, but the actual risk may not be equally applied to all island nations. As noted, the socio-political institutions available to that nation matter: nations with more available resources and planning capacities are finding ways to mitigate and adapt to the current impacts of sea level rise. The same can be said for coastal regions in larger continental areas of the globe. In general, the regions where purposeful management and planning are occurring are developing more resilient policy plans than similar coastal areas where management and planning are not occurring to the same extent. Still, current sea level rise is having impacts in many areas, even where planning is ongoing, suggesting that even excellent plans cannot completely negate the impacts of sea level rise. On the Atlantic seaboard of the United States, for example, "ghost forests" of standing dead trees killed by saltwater intrusion are becoming increasingly common in southern New Jersey, Maryland, Virginia, Louisiana, and North Carolina (U.S. Global Change Research Program 2009; Riggs and Ames 2003). In fact, the entire United States coastal regions are experiencing mostly moderate to severe erosion along coastal areas due to a mix of natural processes (storms and sea level rise) and human activities, including development, dredging, and armoring as populations densities increase along the coastal regions of the United States (U.S. Global Change Research Program 2009).

Societal impacts of sea level rise are evolving. The direction of this evolution depends on both environmental factors and the human response to those environmental factors. Included in the human response is a question of *capacity*, which itself has both social and environmental considerations. A community located on a large continental coastline with a diverse economic base and a strong distributed political system is generally more capable of limiting the societal impacts of sea level rise than a small, geographically and economically isolated island state; the continental coastal community can simply move inland and the socio-economic system in place can more easily absorb this migration. The same cannot be said of the small island state; it does not have the same "capacity" as the continental coastal community in terms of geography and economic tools, thus limiting its ability to mitigate the impacts of sea level rise.

Sea level rise has yet to impact any nation state completely, regardless of its size or resources. However, future worst-case scenario predictions of sea level rise can bring devastating results to coastal regions, wholly submerging low-lying areas, including the entirety of certain island states. This is but one reason why current observations of sea level rise need to be placed into the context of future predictions; observations today offer some evidence of the impacts of sea level rise, but an understanding of potential impacts based on the best science is required in order to make future predictions and then take action based on future predictions. Sound policy comes from understanding both what is happening today as well as what might happen in the future and, when possible, understanding the likelihood of future possibilities.

2.3 Future Predictions

Predicting the future with *absolute* certainty is an all but impossible task. However, a close examination of what is happening today (direct evidence) aligned with a close inspection of what has happened in the past (either direct evidence from past study or indirect evidence from current studies of past events) can help with predictions about the future. Evidence of what is occurring today can present useful information in itself, but usually this information is most useful when compared against a different time. By knowing what is happening today and comparing it to what has occurred in the past, any difference between now and the past can be identified. Indeed, consider the basic observations provided by Charles Keeling when he began to measure the amount of carbon dioxide in the Earth's atmosphere in 1958 (Keeling 1960). The initial measurement in 1958 represented the concentrations observed at that time through direct measurement, thus direct evidence. Repeating this experiment every year using the same methodology, Keeling was able to create a set of data (evidence) that could be used for comparison purposes.

The measurements of 1958 concentrations could be compared to measurements of 1959, 1960, 1961, and so on. As more observations were made and data collected, Keeling and colleagues were able to compare the more recent data (what was happening "today") against the previous data. This comparison allowed for any changes to be noticed between concentrations of carbon dioxide over the years. The more data collected throughout the years, the better the "proof" between comparisons in showing a consistent change (a "trend") over time.

Measurement of the trends (changes in concentrations, if at all, over time) is not the only interesting observation that can be deduced from the data presented; the *rate* of change over time can also be observed. For example, the Keeling data on carbon dioxide concentrations now exists from 1958 up to the present time; well over 50 years of data are available for analysis. Not only can changes in the concentration of carbon dioxide be observed on a year-to-year and decade-to-decade basis, but any changes in the rates of increase can be observed as well. Also, the current information observed can be related to indirect observations through other methods of analysis. For example, cores of glacial ice dated to tens of thousands of years ago contain relative concentrations of what was in the atmosphere when the ice froze (Schlesinger 1997). These cores provide evidence of carbon dioxide concentrations of the Earth in the distant past. Historic concentrations can be compared to current concentrations to gain further evidence about the trend of carbon dioxide concentrations in our atmosphere over time. In addition, other evidence about the climate of the Earth (was it warmer or colder) can be compared to the carbon dioxide concentrations found in glacial ice and more information can be correlated to better understand the effects of increasing concentrations of carbon dioxide in our atmosphere.

Rates of increase are an important observation to note because they can help predict future concentrations with the basic assumption that past trends are indicators of future patterns. This kind of information can be highly informative in an area such as sea level rise where the future expansion of the sea inland is a vital consideration for planning purposes. Basing future predictions on comparing current and past observations limits the amount of "guessing" involved in making future predictions. Below, the best estimates of future predictions are made based on current and past observations, both direct and indirect, relative to sea level rise and associated phenomena. The future predictions identified are categorized into *physical*, *ecological*, and *societal* impacts to reinforce the major categories of impacts for later reference under the policy and legal sections of this text.

2.3.1 Physical Impacts of Future Sea Level Rise

The physical impacts of sea level rise are primarily expressed in predictions about the extent of sea level rise in the future. The reason this is so is because the extent of sea level rise provides an upper limit from which related impacts may be assessed. For example, the certainty of sea levels over the next 50 years allows for precise calculations based on local conditions; the impact of a 6-foot rise in sea level within 50 years would provide an inundation of 30 feet of land in an area where the slope of the nearshore is at a 1-to-5 ratio (for every 1 foot of rise, the sea moves landward 5 feet). Planners could then use this upper limit of a 30-foot migration of the shore landward as the basis for assessing coastal impacts. In addition, this 30-foot approach would provide a basis for determining related physical impacts to the coastal zone, including the extent of flooding of coastal areas, intrusion into nearshore groundwater, and other associated impacts. However, such certainty does not exist. In fact, there are a number of factors that influence the extent of sea level rise over time; depending on the extent of sea level rise, other physical impacts may be present. The current knowledge of the future extent of sea level rise is described below in order to provide a sense of the range of potential impacts.

As noted earlier in this chapter, sea levels have risen at an average rate of 1.8 millimeters per year between 1961 and 2003, with the average annual rate of rise increasing to 3.1 millimeters per year between 1993 and 2003. This rise in sea level has occurred during a time when global ice cover has also been observed to be lessening and the world's oceans have been observed to be getting warmer (IPCC 2007). This information has provided a sense of what has been occurring in the past as well as the current state of sea level rise based on the physical impacts, including recorded increases in sea level, observed and recorded decreases in global ice cover, and recorded warming of average sea surface temperatures. Based on these observations, future predictions can help provide a relative probability as to the future extent of physical impacts, primarily by identifying the likelihood of future increases in sea level rise.

Future predictions of sea level rise are based primarily on models that take into account past and current rates of sea level rise, combine associated factors such as concentrations of greenhouse gases in the atmosphere and feedback mechanisms that reinforce a warming planet, and then extrapolate this information to future trends (IPCC 2007). The resulting estimates of future sea level rise are generally presented in categorical estimates—for example a low, medium, or high range of estimates. Low ranges represent the most conservative estimate of sea level rise based on an assumption that observed factors will be limited in their impact on forces that would attenuate sea level rise. High ranges represent a more liberal assumption where precipitating factors of sea level rise are unabated, causing maximum potential sea level rise. Sometimes, high ranges include a "worst-case" scenario representing the extreme of sea level rise where, for example, major portions of glacial ice are melted absorbing the volume of that ice into the oceans. Understanding the assumptions behind the ranges given for future sea level rise is important when policy decisions are being contemplated and different ranges are being observed to inform those policy decisions.

So what ranges exist for future sea level rise? The answer is somewhat dependent on which source of information is considered to answer the question. The Intergovernmental Panel on Climate Change (IPCC), arguably the most reputable source of information on sea level rise predictions because of the exhaustive and iterative process employed by this group in determining a range of sea level rise, initially predicted sea level rise of approximately 1 meter by 2100 (IPCC 2001). In a more recent updated report, the estimate of average global sea level rise was reduced to 0.6 meters by 2100 (IPCC 2007). The change in estimates between the third (2001) and fourth (2007) assessment are based primarily on additional data (measurements) obtained between assessments as well as changes made to the methodology (assumptions) employed by the computer model when extrapolating the data to predict future sea level rise.

Other estimates have been done on the future extent of sea level rise, including estimates that contain higher resolutions on the amount of average rise that may likely be observed over shorter time periods. Hemming et al. (2007) predict a sea level rise between 8 centimeters (low range) and 13 centimeters (high range) by 2030; between 17 centimeters and 29 centimeters by 2050; and between 35 centimeters and 82 centimeters (0.82 meters) by 2100. Most of these estimates are based on average predictions that discount the worst-case scenarios where a high-impact, low-probability event occurs such as the complete melting of major glacial ice sheets. For example, the melting of either the Greenland or West Antarctica ice sheets would result in an average global sea level rise of 20 feet (over 6.0 meters), while the melting of the East Antarctica ice sheet would result in an average global sea level rise of approximately 200 feet (over 60 meters) (U.S. Global Change Research Program 2009). If either of these high-impact events were to occur, it would dramatically change the question of the extent of physical impacts on the coastal zone.

Whatever the future extent of sea level rise, the effect on different areas of the Earth will not be equal. Regional variations in the gravitational field of the Earth based on shape and the unequal distribution of the mass of the Earth will result in varying degrees of sea level rise, depending on location. For example, the Pacific and Atlantic coasts of North America will generally experience higher sea level rises than other coastal regions, and this will be exacerbated by the collapse of the West Antarctic Ice Sheet as weight distributions between northern and southern hemispheres will be heightened (Bamber et al. 2009). The take-home message here is that the extent of sea level rise is very much dependent on specific variables; some sea level rise is essentially guaranteed, but the physical extent of sea level rise is an open question, and the relative impacts on certain regions of the Earth can be disproportionately impacted by specific high-impact events.

One interesting point in thinking about the future extent of sea level rise is the substantial uncertainty in the current projections. While there is little to no uncertainty that *some* sea level rise will occur in the future, there is uncertainty about the *extent* of that rise. For example, the IPCC estimates have purposefully left out high-impact events in deriving the estimates, particularly the melting of glacial ice sheets, because of the relative uncertainty around these events. However, recent research is beginning to shed light on the actual probability of the melting of these glaciers. Observed loss of glacial ice on the Greenland Ice Sheet over the past decade is twice as much as was estimated by the IPCC Third Assessment in 2001 (Ringot and Kanagaratnam 2006; Dasgupta et al. 2007). Even greater total volume losses were recorded in Antarctica and particularly on the West Antarctica Ice Sheet; if this ice sheet were to simply break away from the continent and slip into the ocean, the displacement alone would raise average sea levels by 3.2 meters in one estimate (Bamber et al. 2009) and approximately 5.0 to 6.0 meters in a separate estimate (Tol et al. 2006).

Uncertainty as to the extent of sea level rise creates uncertainty in policy directions. Obviously, if the worst-case projections of sea level rise were incorporated into policy decisions, immediate and substantial retreat from the coast would be the only logical solution. However, categorical retreat is an unappealing, and politically untenable, option in most coastal management decisions. This fact alone may create incentives for policy makers to discount science that increases the likelihood of a high-impact event, such as a large glacial melt, occurring. At a minimum, decision makers need to be aware of how scientific consensus is developing around the question of the extent of sea level rise because this information will be critical in establishing policy directions to deal with the impacts of such a rise in coastal areas. Adjustments to moderate sea level rise, such as the creation of coastal barriers, may be insufficient to prevent unanticipated rises in the sea level, and such actions today may unreasonably create incentives to stay and develop in otherwise sensitive coastal regions. The role that sea level rise projections play in management choices is further discussed in Chapter 3 of this text.

2.3.2 Ecological Impacts of Future Sea Level Rise

The ecological impacts of sea level rise in the future will depend in large part on two factors: (1) the extent of physical sea level rise (how much the water rises) and (2) the physical features of the coastline in question. The extent of physical sea level rise will significantly impact the changes observed along a particular coastal region; a sea level rise of 1 meter will create impacts different from a sea level rise of 6 meters. The physical features of the coastline will also matter when it comes to determining ecological impacts of sea level rise. A coastline such as the White Cliffs of Dover in England, which reaches upward of 110 meters, will generally experience less of an ecological impact from sea level rise, regardless whether the rise is 1 or 6 meters in height; the cliffs act as a natural barrier with a stark transition between water and land that lacks any substantial coastal zone features as described earlier in this chapter.* Similar to the White Cliffs, human-built sea

* This is not to suggest there would be no ecological impacts from a sea level rise on areas of the White Cliffs themselves. Current exposed areas of the cliffs that would be submerged after sea level rise reduce the amount of cliff area that may be utilized by marine and bird species for habitat purposes. Also, the increase in sea level brings new tidal dynamics to areas of the cliff below and above the new tidal ranges.

walls and barriers against the sea create an area where ecological impacts will likely be minimized because preexisting barriers abutting the ocean have already removed much of the intertidal zone where a larger range of habitat mosaics would otherwise occur (Kelly 2000).

Putting aside for the moment the question of the extent of sea level rise, current evidence suggests some continued rise will undoubtedly occur for the foreseeable future. This sea level rise will present itself gradually in most scenarios, with the potential for discrete high-impact events to occur intermittently, causing possible punctuations in this otherwise gradual progression forecast (IPCC 2007). Gradual manifestations will include changes to the coastal landscape, primarily through coastal erosion and subsidence. In addition, there will be changes in the hydrology of the region, as well as changes in water temperature and chemistry (salinity) with ultimate inundation of coastal areas (Beever 2009). Changes in the morphology, hydrology, and chemistry of coastal areas can impact species diversity by modifying current habitat conditions. While ranges of certain marines species will be expanded through events such as inundation of dry land and increases in salinity levels of coastal waters, displacement of plant and animal communities along coastal areas will occur in the near- to mid-term due to erosion and wetland habitat loss (U.S. Global Change Research Program 2009).

High-impact events can cause abrupt changes in the morphology and ecology of coastal systems. Increased storm occurrence and intensity along coastal areas will likely occur because of the inland reach of rising coastal waters as well as the associated warming of those waters (U.S. Global Change Research Program 2009). Flooding and inundations from individual storm events will occur more frequently as sea level rises, as will sea surges bringing high-salinity content water to near-coastal fresh water (surface and groundwater) and rising water tables (Nicholls and Tolls 2006, p. 1075). The effect on ecosystems can be substantial because of the dynamic nature of storm events and the attendant reduction in time to respond to the kinds of changes brought on by dynamic events. In an ecologically ideal situation, sea level rises at an incremental rate over a long period of time. The slow rate causes a relatively slow pattern of change along coastal zone features; salinity rises slowly in coastal wetlands and watersheds; and habitats of nearshore uplands are slowly inundated with water as the transition from upland to intertidal zone occurs over an expanded period of time. Because the rate of change is relatively slow under this scenario, species have an increased chance to adapt to the changes. Abrupt changes where these kinds of processes happen very quickly leave little time to adapt, placing significant pressure on existing species (Oliver-Smith 2004; Nicholls and Hoozemans 2005). Thus, the ecological impacts of high-impact rapid events can be greater than a slow and steady rate of change over an expanded period of time.

The relationship between the rate of sea level change and the impact on the ecology of an area deserves special attention because without knowing the precise rate of sea level rise in the future, and particularly without knowing the likelihood of high-impact events in the future, the question of ecological impacts on future sea level rise is very much an open question. Because the ecological value of coastal systems is so high (Costanza et al. 1997), understanding how the resiliency of ecological systems are impacted by different event scenarios can help inform decision makers of the potential risks involved in the different assessments. This can be particularly useful where ongoing scientific research is shedding light on the likelihood of high-impact events, such as a large glacial break or melt, when decision makers are considering options today to mitigate and adapt to an unknown sea level in the future.

Recall that impacts to ecological values are documented earlier in this chapter. Figure 2.5 presents an example of a generic coastal system containing many of the physical features (zones) generally observed along a coast. These several zones provide important habitat for a variety of species that depend on the particular characteristics of each zone for their survival and prosperity. For example, certain grasses and shrubs that occur near coastal waters in areas where fresh and salt water meet (sometimes referred to as *coastal marshes*) survive only within specific salinity zones that are diluted from the higher concentrations seen in coastal ocean waters. Saltwater inundation into these marsh areas via sea level rise increases the ambient salinity levels and prevents these lower salinity plants from being able to survive in the new, higher-salinity environment (Schlesinger 1997). When sea level rise is occurring at a consistent and stable rate over a longer period of time, the grasses and shrubs can better adapt to the rise in salinity because it is based on a slow progression; the plants have a better chance of migrating upland because the slower progression allows for enough time to adapt to the changing environment. Also, there is sufficient environment just outside the zone of direct influence available to the plants so they have a place to transfer seeds for the successful migration. When sea level rise is abrupt, there is less time for these plants to accommodate the rapid change in salinity levels. A fast inundation of sea water into a marsh area that fundamentally changes the chemistry of the water can prohibit the plant species from adapting to the change—there simply is no time for adaptations to occur for the species.

Species' capacity to adapt to changes in their environment is well documented in the Earth's history through a close examination of the geologic record. Most evidence shows that abrupt changes in environmental conditions can dramatically influence species survival, with the majority of mass species extinctions noted throughout geologic history correlated to changes in the prevailing environmental conditions (Fastowsky and Sheehan 2005). A stark but generally well-known example is the prevailing theory of the "K-T"—Cretaceous-Tertiary (more recently referred to as the "K-Pg"—Cretaceous-Paleogene) extinction event about 65 million years ago. This mass extinction event, which killed off many of the non-avian large dinosaurs, is believed to have occurred through a number of catastrophic events, including a massive asteroid impact, that altered the environmental conditions on the Earth in such a way that brought about a high level of extinction over a short period of geologic time when compared to background extinction rates (Fastowsky and Sheehan 2005). Like observed extinctions in our geologic record, the extent and rate of future sea level rise will undoubtedly have an impact on the extinction rates of species that utilize the coastal zone in their respective life cycles; the more pronounced the environmental change based on rate and extent of change, the greater the likelihood of larger-scale ecological impacts.

Human interactions with the coastal environment can also have ecological impacts that are both driven and intensified by sea level rise. Many developed coastal areas are buttressed against incoming tides through armoring techniques that prevent the intrusion of the sea past a certain point (Titus 1998). Sea walls, one example of armoring, provide a stark contrast between the sea and land, often allowing the water to come into direct contact with a wall structure, depending on the distance between the wall and the ocean. Because the tides generally impact sea walls directly,* intertidal zones do not develop in these land–sea intersections limiting the types of environments, and thus species, found in these buttressed coastal regions. Collectively, the kinds of plants and animals found in these areas contrast with similar undeveloped coastal regions where intertidal zones are maintained (Titus 1998).

The proximity of human habitation near coastal areas can also have ecological impacts without armoring against the sea. Alteration of coastal landscapes to support residential, agricultural, and other forms of development influence coastal species. Phragmites, a species of large perennial grass, is often found in developed coastal areas replacing local coastal plant species. The large grass alters preexisting habitat, thus changing the landscape mosaic, which in-turn alters the types, concentrations, and distributions of species found in certain coastal regions (Meyerson 1999). The development of residential homes abutting the coast also impacts ecological functions of the coastal zone; for example, increased use of fertilizer to maintain residential lawns creates nutrient enrichment of coastal waters that can lead to low oxygen conditions and toxic algal blooms, both of which impact coastal ecosystems (Schlesinger 1997, p. 224–260). The ecological impacts can intensify where sea level rise brings the interactions between human development and coastal processes closer together. Inundation of coastal waters can overrun human coastal developments, causing septic systems to fail and discharge untreated waste directly into coastal waters (Titus 1998). These kinds of interactions collectively produce ecological impacts that are intensified through sea level rise.

The connection between human development and ecological impacts presents a particular set of problems for the policy maker. Development is a hallmark of human civilization, but so is the protection of citizens by their government from known and potential dangers. The unknown extent of sea level rise along the coast

* The actual impact between a sea wall and the ocean depends on how the wall is constructed and placed in position relative to the coastal area. A wall placed further away from the sea may allow for some intertidal zone area to develop. However, continued sea level rise would generally bring the sea closer to the wall over time. This is discussed in further detail in the policy options section of this text.

presents a set of dangers to the public that must be internalized in coastal planning. For example, does it make sense to allow for new development in a region of the coast that might be particularly susceptible to sea level rise? If development is contemplated, should it include the proactive placement of barriers such as sea walls to protect against the incoming sea? What role should ecological impacts play in making a decision on whether or not to develop and armor in a particular region? And what about existing developed coastal regions: what role should sea level rise predictions play in how these areas are managed? Does it make sense to proactively engage in policies that favor retreat from the coastal zone rather than staying and armoring against the sea? Again, what role do ecological values play in making these kinds of decisions? These are some of the questions that are presented later in this text as we discuss policy options. However, understanding the impacts that are at stake in making these decisions is a critical part in fully understanding coastal planning in a world where the seas are rising.

In summary, the future ecological impacts of sea level rise contain elements of both certainty and uncertainty. Certainty lies in the fact that sea level rise is occurring, and depending on the rate and extent of future sea level rise, there will be impacts to the ecological health of coastal communities. There is also near certainty in the proposition that continued sea level rise will have continued ecological impacts on coastal systems. The extent of the ecological impact depends on a number of factors, including the rate and ultimate heights of sea level rise that actually occurs in the future; greater sea level rise that occurs at faster rates will tend to have a greater impact on coastal ecology than less sea level rise attenuated over a longer period of time. The uncertainty of how much sea level rise will occur in the future makes it difficult to define the extent of ecological impacts. However, the worst-case scenarios of sea level rise provide bounded upper-limits that can be used to determine the potential extent of ecological harm. How this kind of information influences policy directions is a question that is left to the process used in making management decisions in this area.

2.3.3 Societal Impacts of Future Sea Level Rise

Recall that societal impacts from sea level rise are generally determined through two main factors: (1) an assessment of geographic vulnerability of the localized coastal area, and (2) consideration of the population density of the coastal area under consideration (Adger, Vincent, and Vincent 2005). Vulnerability is a measure of both objective environmental factors as well as more human-defined factors that may or may not be present. These human-based factors include awareness of the hazard, intensity, and sophistication of development along coastal regions, and the kinds of public policy institutions established to deal with the risk (Brooks et al. 2005). Collectively, these factors can be compared to the range of sea level rise estimations identified, and by doing so, provide several estimates of potential societal impacts.

Geographic vulnerability can be spatially compared to future estimates of sea level rise. For example, if a conservative future estimate of sea level rise is 0.5 meters over the next 50 years, and a defined coastal area will be totally inundated by the ocean with a 0.25-meter rise in sea level (half the estimated rise), then this coastal area can be defined as geographically vulnerable within a 50-year period because it will be completely submerged under *conservative* estimates of sea level rise within the 50-year time frame. The spatial comparison provides a starting point in determining the potential societal impacts on different coastal areas. This initial analysis can then be supplemented with other considerations such as the population density in the coastal area and the institutional capacity of the coastal area to deal with sea level rise.

Developed coastal areas near current sea levels are acutely vulnerable to sea level rise impacts. Even the most conservative estimates indicate that by 2100 there will be approximately a half-meter average increase in sea levels across the globe. Other factors can influence the impact of sea level rise on a coastal area; for example, much of the City of New Orleans sits below sea level, but it is not under water because of a system of levees that has been built to protect the city from flooding. While technological advances such as the levees in New Orleans can mitigate the impacts of sea level to some degree, powerful storms such as Hurricane Katrina in 2005 provide an upper limit on how much technology and social preparedness can mitigate geographic vulnerability.

Hazard awareness, sophistication (technological, planning, etc.) in coastal development, and established public policy institutions to deal with the risk of sea level rise are mitigating factors when determining vulnerability. Coastal areas that score high on these three factors will tend to have less vulnerability than coastal areas that have a similar composition (same height above sea level, ocean facing, similar gradation inland, etc.) but have less awareness of the coastal hazard, are less sophisticated in their coastal development, or have little to no established public policy institutions that deal with coastal sea level rise. SIDS, as identified previously in Section 2.2.2.2, will tend to have fewer resources available—physically, economically, and politically—to deal with sea level rise. The lack of human and economic capital is heightened by a lack of geographic size, which limits policy options; even if a small island state made preparations to move away from the threat of sea level rise, there is little to no available space for many of them to go (Pelling and Uitto 2001). However, as just noted in the example of Hurricane Katrina and New Orleans, even sophisticated planning and awareness, coupled with the economic capacity to deal with disasters, is no failsafe against rising coastlines and their associated impacts. Higher-intensity storms are of particular concern for coastal areas that are capable of holding back the sea and thus allow civilization to remain in close proximity to coastal hazards.

The other main factor beyond geographic vulnerability is population density: The more people living at or near coastal areas, the greater the risk to those populations from the kinds of effects that coincide with sea level rise. Recent trends in population movements identify net positive coastal migration patterns in many areas of the world. For example, in the United States, there has been a general shift in population to coastal areas, where 52 percent of the nation's total population currently resides in coastal regions of the country, an area that takes up less than 20 percent of the total land area of the continental United States; this amounts to 164 million people living in coastal areas in the United States alone. An additional 14.9 million people are projected to move into coastal regions by 2020, a 9 percent increase over the current coastal population number (U.S. Census 2011).

Associating the two independent trends – increasing sea level rise and increasing populations in coastal areas – identifies the potential for significant societal impacts related to sea level rise. This statement becomes evident when the associated impacts that correlate with sea level rise are identified: increased coastal storm surges; greater subsidence and erosion of coastal land; rising water tables and the intrusion of salt water into groundwater and coastal aquifers. A rising sea also raises the likelihood of dangerous conditions at the coast that make living more difficult, if not impossible. Add in the potential for a high-impact event like a glacier melting and the coast looks downright inhospitable.

Viewing the potential social impacts of sea level rise through the competing lenses of population trends and increasing risks to those populations, coastal planners have a difficult task in balancing the interests of coastal dwellers against the legitimate increasing risks associated with coastal living. This is particularly true where proactive policy development includes placing limits on development prior to the effects of sea level rise being displayed. The potential for public outrage is high when the data suggest a general populace preference for coastal living, but that preference runs counter to sound and rationale policy development that wishes to decrease the risks associated with coastal living. Managers who are forward thinkers can find themselves in a situation where there is little public support among coastal dwellers for policies that restrict the ability to live and function in the coastal environment.

Ecological factors should also be considered when thinking about the societal impact of future sea level rise. As noted above, armoring and other techniques can be applied to limit the vulnerability of a coastal area. However, these techniques impact ecological values that are associated with coastal zones in general, but also cross over into human well-being. For example, armoring can deprive a coastal zone of natural wetland and nearshore marsh areas, including the native sea grasses that provide habitat for commercially important fish species (Titus 1998). The loss of wetland resources through armoring impacts both provisioning and regulating services provided by natural coastal features and thus impacts the human social networks that develop around those services (Kelly 2000). Mitigating vulnerability through armoring techniques can have impacts on ecological resources that are important for direct human use (e.g., habitat for commercial and recreational fish species) as well as indirect values such as storm protection through natural barriers that exist within an unimpeded intertidal zone.* Understanding these impacts helps decision makers contemplate tradeoffs when thinking about mitigation techniques for vulnerability such as armoring.

Another aspect of societal impacts to future sea level rise surrounds the kinds of group values and social decision-making that will emerge as communities consider adaptation choices to sea level rise. The makeup of the coastal community and how the community perceives the coastal zone are important considerations as mitigation and adaptation choices are debated and policy paths chosen. An important factor in this discussion is how the community itself relates to the kinds of rights that exist within the coastal zone. *Property rights* become an important consideration in this process because they can be applied to a variety of social resources that will be impacted by future sea level rise. There are individual property rights in the land of the coastal zone itself; private landowners will want to protect their investments in their coastal property, and this desire will impact policy choices. A policy that requires the retreat from private land due to approaching tides may be challenged by private landowners under existing legal doctrines, depending on how individual property rights toward land are interpreted in the particular coastal area.† Beyond private landowner rights, rights to *resources* within the coastal zone also exist. Policy choices such as armoring that impact important habitat for commercial marine species can also impact the rights of resource users and must be taken into account. Finally, adaptation choices might have to rearrange property right expectations in coastal areas in order to internalize costs associated with the dangers presented through sea level rise and its associated impacts along coastal areas. A number of these property right issues will be further explored in the policy (Chapter 3) and law (Chapter 4) sections of this text.

The adaptive capacity of coastal societies varies with local conditions, and the extent of sea level rise will dictate some options. In some cases, the full inundation of the coastal area will require retreat. In these cases, coastal populations will be forced to move inland when possible, and in the case of certain small low-lying island nations, the population may become sea level rise refugees and have to seek asylum in other nations (Barnett and Campbell 2010). In other cases, partial

* Often, storm protection occurs from a few factors associated with natural coastal zones. First, there is a natural slope where sediment deposits create shallower depths as one approaches the shore from the deeper ocean water. This slope acts as a buffer against tidal storms, causing larger tidal waves to break further out to sea due to the decreasing depth as a wave approaches the shoreline. The normal slope is altered by armoring, allowing the wave to approach much closer to the shoreline before it begins to break. In addition, the natural intertidal zone usually contains a set of features (sandy beach, dunes, marsh, vegetation) that places a larger natural buffer between the ocean and human development. Armoring erodes these areas, allowing the ocean to come within close proximity of human development and thereby increasing the risk of impact from storm events.

† In the United States, private property rights are protected under the Fifth Amendment to the U.S. Constitution from government actions that are deemed to be a "taking" of the private property. This property rights issue is dealt with in detail in the "Legal Considerations" section (Chapter 4) of this text.

inundation may bring about changes to the coastal area that do not require a full retreat from the coast, but the community may need to draw on adaptations to adjust to the changed circumstances. Depending on the resiliency of the coastal community, the ability to adjust to changes in background environmental conditions will vary. Agrarian communities may have to change farming practices due to saltwater intrusion into coastal aquifers and higher salinity of the soil. Geomorphic changes to the coast may also alter traditional practices of the community, such as coastal cropland loss, requiring a change in behavior patterns and practices. The ability of the community to adapt their practices to suit the new coastal environment will be an important factor of the overall adaptability and survivability of the societal unit throughout the evolution of sea level rise.

Finan et al. (2002) have studied adaptation to climate extremes by societies, noting that the ability to adapt to changes in the environment is based on two primary factors: (1) technological capacity and (2) social organization. These two factors combine to determine the extent to which resources will be distributed throughout the community under change, the institutions that are developed to support the changed conditions and thus alterations in behavior patterns of the community, and the extent to which the affected coastal communities are empowered to have a say in the management and distribution of the changed environment (Finan et al. 2002). In developed societies where there is ample land to retreat to, like continental coasts such as the United States of America, a high technological capacity can limit the impact of sea level rise to some extent, while resources can be redistributed and coastal communities can be absorbed into the larger economic and social systems. In smaller island nations with economies that lack scale, the options are diminished and thus the ability of the community to socially organize is an important factor in the adaptive capacity of these particular coastal societies (Finan et al. 2002).

Overall, the trends shown indicate the potential for a substantial future impact to society through continued sea level rise. Humans are migrating to the coast, causing coastal populations to rise. The rise in coastal populations is occurring at a time when sea level rise is being identified as an ongoing phenomenon that will likely only increase for the foreseeable future. Fundamental choices about how to deal with an increasing coastal population exposed to the risks associated with sea level rise need to be made by policy planners and these choices are best done in a proactive way. One problem in being proactive about policy directions is the uncertainty about the future extent of sea level rise. The ranges of future rise predicted are broad, particularly when high-impact events are included in the range analysis. Will sea level rise a few centimeters, 1 meter, 20 meters? There is no sure way to answer this question definitively. Thus, policy decisions about whether to stay at the coast and defend against the incoming tide or to plan for an organized retreat as sea level rises are difficult to make; the uncertainty inherent in knowing what will occur in the future creates difficulties in being proactive. However, the expectations of those living at the coast can help in making these decisions. Community values are an important part of thinking about policy choices as these values underlie acceptable policy choices.

Values that aid in proactive and precautionary policy choices can substantially limit the potential impacts to coastal societies in the future, whereas value expressions that limit policy choices to adequately deal with sea level rise proactively may increase the societal impact regardless of mitigation techniques employed.

Estimates of future sea level rise impact a full range of concerns: physical, ecological, and societal. Even low estimates will have some impact on human well-being in coastal areas; higher estimates can have substantial impacts on coastal resources. Worst-case scenarios are simply catastrophic when applied to the range of impacts they can have on both coastal and inland resources. All of this information leads to a question of what to do today; however, as discussed above, answering this question of what to do is not as easy as it sounds. Both proactive (doing something before sea levels rise fully) and reactive (waiting and seeing) policy stances have their strong points and problems. The scale of understanding these problems will be further discussed in the policy (Chapter 3) and legal (Chapter 4) chapters of this text. Still, understanding the problems creates the foundation upon which those discussions are brought to life. Before moving into the policy section (Chapter 3), some additional discussion of coastal land use features is warranted in order to understand how geomorphology impacts sea level rise. Some observations have already been made about the different kinds of features found at the coastline and how these features might impact the extent of sea level rise as well as the choices made in responding to sea level rise. Providing additional details about the relationship of the structure of the coast to sea level rise can aid in more fully understanding policy options in the context of assessing the varying impacts of sea level rise on different coastal attributes.

2.4 Major Coastal Land Feature Considerations

Major coastal land use features are a critical consideration when understanding the overall science behind sea level rise, including the potential impacts that sea level rise might have on a particular coastal area. For example, a coastal area that is defined by high bluffs situated well above sea level will tend to be better protected against the impacts of sea level rise. This is more so where the makeup of the coastal surface consists of a hard substrate that can better withstand the forces of erosion and associated storm events along coastal areas. Contrast the high bluff, hard substrate coastal landmass just described with a coastal zone that sits near sea level, contains only a slight grade difference between the land and sea boundary, and has a less dense substrate that is more susceptible to the forces of erosion. The physical characteristics that define these two kinds of coastal zones impact the influence sea level rise will have on each type of coastal area; the lower lying coastal area is more vulnerable to sea level rise, while the higher area provides natural barriers and thus protection from both sea level rise itself and the attendant conditions that follow (increased storm intensity, reach, etc.).

Considerations beyond physical proximity to the coastline and density of the substrate underlying the coastline include the location of the coastline, particularly the relative positioning of the coastline in relation to the open ocean; the forces acting on the coastline; and other major features that can influence land–sea interactions such as the presence of barrier islands and similar features along coastal regions. The purpose of this section is to identify major coastal features and discuss the relationship between coastal features generally and sea level rise, identifying where impacts of sea level rise may be either mitigated or exacerbated, depending on the mix of coastal land features observed. The section is divided into several categorical components, starting with the general *geomorphology* (shapes) found in different coastal areas. Geomorphic features are an important first determination in considering sea level rise impact and thus should serve as the beginning point of reviewing coastal land features. The next section deals with *geospatial characteristics* of the coastline; where is the coastline situated specifically, and what forces are acting upon the coastline based on its particular location? This review helps to place the coastline into a location context, helping to show how the specific location of the coastline can influence the impacts of sea level rise. The final categorical component considers the level of *development* around the coastal feature. Human development impacts coastal areas in several ways, as will be discussed, and the intensity of human development along a coastline is a critical feature in understanding interactions between land and sea in developed coastal areas.

2.4.1 Geomorphology Considerations

Geomorphology is, generally speaking, the study of landforms, including the processes that shape land. The goal of geomorphology is to understand the history of land formations (how they formed, what impacts formation) in order to get a better sense of why landforms look the way they do today and also to help in predicting what forces might impact landforms in the future (Chorley, Schumm, and Sugden 1985). By studying the underlying geologic processes of an area and combining this information with the surface processes that impact the landscape, geomorphologists can gain stronger insights into the dynamics that help shape the way land looks and also identify factors (physical, chemical, biological) that may influence landscapes over time. For example, a windswept coastline that is dominated by strong storm surges has the potential to impact the way that coastline looks (its *morphology*). The physical makeup of that coastline matters; if the coastline is made up of softer materials, then it is more likely to be impacted by the forces of wind and wave, resulting in erosion that more easily changes the look of the coastline. Alternatively, if the coastline is made up of mostly hard substrate, then it is physically more resilient to the forces of wind and wave acting upon it. Understanding these basic factors can aid in determining the potential impact of sea level rise on observed coastal landforms.

One way to understand the impact of sea level rise on the geomorphic features of coastlines is to get a sense of the kinds of processes occurring at the coast that have the potential to impact the shape of the coastline. Major processes occurring at the coast that have potential to impact coastal shape include the following: *waves, tides, currents, shoreline weathering, coastal erosion, sediment transport and deposition, organic activity, changes in sea level, coastal hazards* (rip current, storm surges, tsunamis), and *human modifications of coastal processes* including *soft structures* (dredging, beach nourishment, and beach scraping) and *hard structures* (sea walls, jetties, groins). Each of these categorical processes is summarized below. However, prior to reviewing these processes in detail, it is important to place them in the context of this text and that is understanding how knowledge of these different processes is impacted by present and future sea level rise. In a word, these processes should be considered in reference to overall *resiliency.* Resiliency is a measure of both physical resiliency (considering the geomorphology of the coastline) as well as the human-based factors of resiliency, which can be determined in part by looking at the factors of *vulnerability* as described above in the societal impacts to sea level rise materials. Beyond the physical characteristics of the coastline, resiliency is determined by the ability of human beings (and thus human institutions) to both mitigate the impacts of sea level rise as well as adapt to the changes brought about through sea level rise. Human mitigation and adaptation capacity helps to determine overall resiliency of a coastal area in addition to the more "natural" geomorphic factors discussed in this section. After the processes impacting coastlines are summarized below, major geological characteristics will be summarized in order to provide a context for understanding the background issues surrounding resiliency along coastal areas.

2.4.1.1 Major Processes Affecting Geomorphology of the Coastline

There are a number of energy interactions, driven by the heat energy system dynamics discussed earlier in this chapter, that impact coastlines generally and, depending on the physical makeup of the coastline, can impact the way in which the coastline looks and its susceptibility to being physically altered from its present appearance. This section summarizes the major processes that have a capacity to impact coastal features. The actual capacity of each process to impact coastal features depends on other factors, including what the coastline is made up of (discussed immediately below), where the coastline is located in relation to these processes (discussed in Section 2.4.2), and the human capacity to mitigate the impacts through different techniques including the alteration of the coastline to increase physical resiliency to these processes discussed in Section 2.4.3.

Wave action is one of the processes that acts upon coastlines. Waves are characterized by their length, height, velocity, and period or interval of time between successive wave peaks passing a given point. Collectively, these several properties of waves combine to influence the generation, quantity, intensity, and thus impact of waves on coastlines. However, it is the relationships between these wave properties that ultimately influence the overall impact of waves. Factors that impact the relationship of wave characteristics include what is generating the wave (normal gravity and wind forces versus an acute intense cause such as a hurricane or other intense storm) and the background environment in which the wave exists. Even normal wind patterns can act upon the overall intensity of the wave that is observed at the coast. For example, the energy transfer of the wind to the surface water is cumulative, meaning the more time and distance the wind has to transfer energy to the wave the greater the wave generation and impact along the coast; the more powerful waves are produced by sustained winds blowing in the same direction over a long distance. Thus, the unobstructed distance between a water body and land is a factor in wave impact because it influences the amount of distance available for a wind-generated wave to develop: The greater the uninterrupted distance, the greater the wave generated (Schlesinger 1997, p. 291–340).

Wind energy can also be intensified by other forces, including the development of major storms such as hurricanes that often form in the central regions of major ocean bodies (Schlesinger 1997, p. 77–80, 292–298). These strong wind patterns often create intense energy transfer with the surface of the ocean causing storm surges where a large swath of water is pulled onto coastal land through the force of the storm itself. In such cases, less distance is required to develop strong wave patterns. Beyond wind energy, waves are also generated by gravitational force, although these waves are really part of the large movement of oceanic water within tidal ranges. Finally, displacement of the ocean floor through tectonic movements (earthquakes) can also cause waves of intense size as the water above the ocean floor is displaced and moved onto land.

Tides are the result of gravitational forces acting on the ocean from nearby planets. The moon and sun are the closest large planetary bodies to the Earth so they provide the greatest influence over the tidal ranges observed. The moon exerts a larger gravitational influence on the tides because, although it is much smaller and less dense than the sun, it is closer to the Earth and thus its proximity is the major factor in influencing the Earth's tidal ranges. Tides are important because they represent a constant movement of water over time and this impacts coastal considerations such as rates of erosion. In addition, tidal ranges are important because they influence the ecosystems that develop in a region based on the extent of the range; the Bay of Fundy on the southeast coast of Canada in the Atlantic Ocean has a tidal range of more than 50 feet of depth, resulting in a massive intertidal zone between low and high tide. During the low tide, a wide swath of land is exposed that is later submerged underwater at high tide. The ecosystem developed in this region evolves around that tidal range and changes to the tidal range ultimately impact the existing ecosystem.

Currents are the horizontal movement of water, and the direction in which water flows can impact the dynamics of the coastal area under consideration. Wave energy reflected back into the ocean from the shore is referred to as cross-shore current (sometimes undertow), and such currents have the capacity to drag sediment from the shoreline out to sea, depending on the bottom type of the shoreline as well as the amount of wave energy at the particular location. Some of the water flowing into the shore that is not reflected back out to sea moves parallel to the shoreline; these types of current are normally referred to as *longshore currents*. A longshore current can be powerful, depending on the amount of energy dispersed transporting bottom sediment. Often, barrier beaches or sand bars are the result of a strong longshore current that cuts between the immediate coastal shoreline and the observed sand bar in the water; the current has essentially dug out the sand that would normally exist between the sand bar and the shore, and sometimes the current has acted to build up the sand bar just offshore as it deposits the sandy bottom it has picked up along the shoreline. Because currents are impacted by the amount of energy creating the current, increases in energy will ultimately lead to increases in current potential and impact.

Shoreline weathering represents the process of physical and chemical interactions along a coastal region that results in impacts on the coast with the extent of coastal impacts dependent on the features of the coastline and the energy involved in the weathering process (Summerfield 1991). The process of weathering occurs when the ocean water wears against the intersection between land and sea. The makeup of the coastal landmass matters because it influences the extent of the weathering process; for example, a rocky coastline may be more capable of absorbing the weathering processes than a less dense coastline composition. The chemistry of the water matters as the relative salinity and pH of the ocean water can impact the erosion rates of the coastline depending on the chemical composition of the coastal features; the mineralogy of the particular rocks found along the coast can make them either susceptible or resilient to the chemical weathering effects of the ocean water. The intensity of the tidal flux and extent of the tidal range also matter; a strong tidal range will carry greater force against the coastline, while a large tidal range will expose portions of the coastline to extended periods of drying and wetting, which can exacerbate weathering processes. Finally, temperature can play a role as high latitude coasts may experience low temperatures during cold weather events and time periods, which can intensify weathering through intermittent periods of freezing and thawing. However, the extent of freezing and thawing is somewhat dependent on the relative salinity of the sea water; lower salinity coastal areas, generally occurring in coastal inlets where fresh and salt water mix, will tend to have greater freezing potential and thus cold temperature impact on coastal features.*

Coastal erosion is the process of coastline change brought about through various forces, including weather and wave activities. Weather processes can drive coastline

* High-salinity water contains dissolved solids that come out of solution as salt water freezes. This produces a low-density ice that does not create a strong penetrating effect of fresh-water ice on hard coastal features such as rocks. The lower the salinity of the coastal water, the greater the density of the frozen water and thus the greater potential for impacting hard substrate structures along the coastline.

changes in ways that were identified above. For example, seasonal variations in higher latitudes can bring about ice accumulation at the coast that can stimulate coastal erosion. Also, cyclical weather patterns in certain regions can lead to strong storm occurrences (hurricanes) that can bring about and intensify background rates of coastal erosion. Wave action can also influence coastal erosion in various ways, depending on factors that influence the formation of the wave pattern as well as the geomorphic condition of the coastline itself. Waves that crest and break at the coastline carry a greater intensity of energy with them and have the potential for impacts along coastal regions where the energy is dispersed in short but powerful sequences. The general shape of the coastline influences this process as stronger breaks are found near coastal areas where the seafloor abruptly shifts from a greater depth to a shallow depth near the shore over a short distance. The rapid decrease in depth allows for the cresting of larger waves near the coastline, resulting in a break that captures air (sometimes referred to as a *plunging breaker*) and pushes the air and weighted water down toward the coastline. Coastlines that are defined by high bluffs generally do not have a sharp decrease in depth approaching the coast and therefore waves do not crest and break allowing the energy of the wave to be reflected from the coastal body with less intense energy dispersal. Other factors, including watershed interactions with coastal embayment areas, can also influence coastal erosion patterns, for example, depending on the volume and intensity of river water discharge into an estuarine environment during seasonal variations and storm events (Schlesinger 1997, p. 261–290).

Sediment transport and deposition is a phenomenon that occurs in coastal areas and depends on several factors, including the makeup and composition of coastal landforms, the physical characteristics and dynamics of the coastal zone, and the dynamics that occur offshore in the waters affecting coastal conditions. Recall that wave action and energy exchanges can result in the movement of bottom sediments along the coast; refractions of wave energy along the coastline can result in longshore currents that both erode and carry sediments parallel to coastlines. The bottom type of the coast can either accelerate or mitigate sediment transport and deposition; coastal seafloors consisting of larger rocks and boulders will mitigate the impacts of sediment transport, while smaller sandy and granular bottom types can accelerate the impacts of sediment transport depending on the intensity of the energy transferred during wave and current activity. The physical makeup of the coastal landforms can also impact sediment transport where softer compositions such as sandstone can lead to higher rates of erosion, which can contribute to sediment accumulation in the nearshore. Denser compositions such as granite are less prone to erosion and thus there is less land-derived material available for transport and deposition.

Overall, the greatest contribution of sediment transport comes from rivers. Riverine effluent, on a global basis, contributes about 100 times as much sediment as marine erosion (Schlesinger 1997, p. 261–290). Rivers carry large amounts of sediment downstream and many of the largest rivers displace their contents directly into marine estuaries and bays where the sediment then becomes available to the forces of waves and currents acting as identified and explained above. Sediment deposition from river sources is not equal in terms of impact around the globe. The greatest impact of river-based deposition of sediment is observed in the tropical latitudes near the equator where low energy patterns cause little marine transport through wave action (Schlesinger 1997, p. 291–298). Higher latitudes experience greater wave energy dynamics along coastal areas but rivers are still a significant source of sediment for transport and deposition in the marine environment. Human activities that remove this sediment (dredging bays or damming the river upstream trapping the sediment) impact the amount of sediment that reaches marine coastal areas and is available for deposition and transport. In addition, human structures that impede longshore current transport of sediments (jetties and groins) for example, by creating a perpendicular structure to the parallel transport of sediments along the coast limit the extent of sediment transport allowing material to accumulate in areas where the structure has been erected and thus depriving down-current locations of sediment deposition that would occur without the structure.

Organic activity represents either erosion or creation of marine land mass induced by plants and animals in coastal areas. Certain marine organisms that live along intertidal zones attach to substrate surfaces and bore into them to seek shelter or otherwise create a holdfast that allows them to withstand the forces of wave and currents along this dynamic environment. The attachment and boring into marine substrates can cause accelerated erosion, particularly for sedimentary rocks and similar softer substrates. Other marine animals, such as coral, build elaborate structures as part of their life cycle that, over time, develop into large-scale communities with substantial substrate footprints. Coral reefs provide an important habitat for a diverse number of marine species, but they also provide a barrier to dissipate waves and other energy-derived phenomena that might otherwise cause sediment deposition. In this way, marine animals can help to stabilize sediment in the areas where they exist. Finally, marine plant communities dominate many near-coastal areas, helping to develop salt-marsh communities and associated wetlands within and near intertidal zones. The plant communities trap sediment in their root systems, thus impacting the deposition of sediment across the community. In addition, a large community can buffer the impacts of sediment transport, allowing less sediment to be moved from the area in which the community exists (Schlesinger 1997).

Changes in sea level impact the morphology of the coastline because the relative position and height of the sea in relationship to the land determines the location and particular shape of the coastline. Climate change can impact sea level rise as described earlier in this text by causing glacial melt water and thermal expansion to combine, thereby raising the average level of the sea. Raising the sea level will impact current coastal features by inundating areas of the coastline that are now dry. Some of this rise may be mitigated in landmasses where glacial melting results in a "crustal rebound" of the land from the loss of the weight associated with the glacier (Summerfield 1991). Continued tectonic processes can also influence the level of the sea in relation to a particular coastal area. Variations in relative sea level may also result from geodetic changes such as fluctuations in the angular velocity of the Earth or polar drift. Impacts of changes in sea level can include coastal flooding, waterlogged soil, inundation of coastal wetlands, destruction of coastal communities, salinization of coastal aquifers, and ecosystem degradation (Schlesinger 1997).

Coastal hazards represent natural processes that impact rates of erosion, sediment deposition and transport, and other contributors to changes in coastal morphology. The term "coastal hazard" is human centric and represents processes that impact human well-being at the coast because of the intensity and thus associated dangers that often accompany these various phenomena. However, these processes have been built into the natural environment over a long period of adaptation so that many of them are necessary to supply certain habitats with needed resources. For example, storm events provide shallow marine and estuarine ecosystems with critical sediment replenishment, a process that might become even more important for these ecosystems in the face of increasing sea level rise. Coastal hazards described here include *rip currents*, *storm surges*, and *tsunamis*.

Rip currents are a phenomenon that can be explained through the general laws of physics. A rip current occurs as water builds up along a coastal area due to repeated wave activity. As waves continually break along the coastline, water volume builds up so there is a slight increase in water level from the break of the waves landward than there is from the break seaward. The difference in water levels creates instability where the volume of water near the coast that is seeking to be removed from the coastal area is greater than the volume of water between the two areas. A current near the coastline forms at a low point underwater that is stronger in force than the background current of water retreating from the coast after a wave break. The swift-moving rip current carries the excess water back out into the ocean, providing not only a danger to humans but also a strong transport mechanism for sediment that has the potential to alter the coastal landscape (Summerfield 1991).

Storm surges are extraordinarily high water levels (compared to average background levels) and are caused by a combined effect of low atmospheric pressure and high wind speeds (Nichols and Tol 2006). A storm surge can impact the morphology of the coastline by bringing large quantities of sea water inland. Low-lying areas along the coast can be inundated for prolonged periods after the storm surge event. Because storm surges coincide with storm events (gales, hurricanes, etc.), they also bring the potential for high rates of erosion over short periods of time: Both the surge and the rainfall from the storm can contribute to erosion. The effect of a storm surge can be heightened if it occurs during a high-tide event because the surge is bolstered by higher average sea levels, thus heightening the extent and impact of the surge. The association of the storm event itself along with tidal maximums can create extensive damage to coastal areas, altering the coastline's appearance, irreparably in some cases (Pinet 1992). Storm surges are most common along coastal areas that are located in mid-latitude regions of the world where hurricane events (or "seasons") are common. The main capacity for geomorphic impact comes from the fact that storm surges provide water intrusion into land areas normally untouched by background wave activity; many of these areas are particularly sensitive to the saline water that accompanies the surge.

Tsunamis are large ocean waves that are generated through seismic activity that can include the shifting of submerged geologic plates causing a displacement of water. Other causes of tsunamis are also geologic activity, particularly earthquakes and volcanic eruptions. Tsunamis that begin in the middle of a large deep ocean often are unseen because the wave height is only a few inches in comparison to its very long length. Only when the wave approaches a landmass does the length of the wave begin to shorten and the full height become realized. Some tsunami waves can exceed 30 feet in height. These high-impact, low-probability events can cause substantial morphological disruption in the coastal zone through mass inundation of sea water inland and the attendant erosion that can result (Pinet 1992).

Human modifications of coastal processes are also responsible for changing the shape and structure of coastlines. As indicated earlier, more than 50 percent of the U.S. population lives within 50 miles of a shoreline. Development around coastal areas includes reinforcing structures such as seawalls, jetties, and groins. As discussed earlier, these structures impact the normal flow of sediment transport and deposition (jetties and groins), and also alter the natural transition zones between the sea and land (seawalls). Collectively, these human modifications result in alterations of the coastal habitat itself, sometimes dramatically.

As humans settle coastlines, they create commitments to the area through the expenditure of resources in settling and developing a coastal region. The commitment to the area reinforces choices that support stabilization efforts to protect developments. Already mentioned are armoring techniques to hold back the sea such as seawalls. Other human actions aimed at protecting developments along the coast include the creation and stabilization of inlet areas to establish protected bays for navigation, commerce, and recreation. Beach nourishment is another technique employed to secure a safe distance between the sea and coastal properties; essentially, sand or some other "nourishing" substrate is brought to a coastal area and placed between a line of property and the ocean to help secure the area and hold back the tide. Sometimes sand dunes are also created to help mitigate the effects of erosion in a particular coastal region. Also, wetlands are sometimes drained or filled in order to allow for continued development. Dredging of shipping lanes to preserve a minimum depth is also a common practice in human inhabited coastal areas. Collectively, all these activities contribute to alterations of the coastline, sometimes preserving land and other times accelerating erosion processes. Those structures that are "soft" in nature (nourishment projects, dredging, reshaping coastal dunes, and sandy areas) tend to have less impact over time than "hard" structures such as seawalls and bulkheads that tend to have a more lasting impact over a longer period of time.

This concludes the summary of the processes that affect the geomorphology of the coastline. What has been summarized is a set of energy flows that cause a variety of conditions, which collectively have the potential to define the shape of coastal areas. Wind can erode the physical structures of coastlines and also move water in ways that impact coastal and submarine features. The waves generated through wind, gravity, and seismic activity can be forces of change in themselves. So too can the tides. Currents move sediment, while marine plants and animals can aid in keeping sediment in one location. Changes in sea level impact coastal processes, including rates of erosion and the intrusion of sea water landward. Humans engage in a variety of coast altering activities. However, the ultimate capacity of these processes to alter coastal formations (with the exception of humans) is highly dependent on other factors, most particularly the major geological characteristics of the coastline. As hinted above, the makeup of the shoreline matters when we consider the impact of erosive processes on coastal areas. A hard substrate will weather less than a soft substrate. Chemical reactions will be impeded if the material of the coastline is nonreactive to the sea water and other chemical constituents. In short, geologic characteristics matter when trying to fully understand the impact of geomorphology on sea level rise.

2.4.1.2 Major Geological Characteristics of the Coastline

Geomorphology—the way the coastline looks—is an important consideration in understanding the potential impacts of sea level rise. Certain features of the coastline will be more resilient than others based on the processes occurring along the coast, what the coastline is composed of structurally, where the coastline exists in relation to the ocean, and also how human development impacts coastal features. This section focuses on what the coastline looks like from a structural standpoint. Coastlines that are made of hard substances and include high bluffs situated well above the physical processes of the ocean will tend to have more resilience to sea level rise than coastal areas that are made of softer substances where the landmass is near sea level. The goal in this section is to summarize the major features that can be found in coastal areas. The several features identified here can be analyzed in light of the different processes that occur along a particular coastline, the location of the coastline, and human interactions in order to gain a greater understanding of how major coastal land features impact sea level rise considerations.

Recall Figure 2.5, which identifies the general components of a coastal zone. The features present in the coastal zone often include the following:

- A water line that distinguishes between land and sea
- An intertidal zone that represents the fluctuation of the land–sea interface based on tidal range
- An area above the influence of the tide (sometimes a sandy beach)
- A vegetation line that represents the landward extent of the coastal area

The general characterization of coastal zone features in Figure 2.5 does not represent *all* varieties of coastal features found in a particular coastal zone area. There are often variants in how a coastal zone looks that represent the local conditions experienced in the particular area; a close inspection of the physical environment observed can give clues to the conditions of the area. For example, exposed rocky shorelines lacking sandy beach areas are often caused by high-energy impacts, suggesting there is substantial wind and wave energy that batters the coastline removing smaller particles that would otherwise help create a sandy beach area. The appearance of large rocks also can tell the observer something about the relative density of the substrate along the coast: Larger rocks indicate a harder substance along the coast. In contrast, a sandy beach area may represent a coastal environment dominated by less-intense energy flows with a less-dense substrate capable of being eroded into finer particles through the forces of erosion.* Areas with substantial sand buildup may be endpoints for currents that deposit bottom materials from other areas giving further clues to the interactions of energy processes and what is actually observed. In sum, a variety of factors help to define coastal environments, including wave energy, tidal range, sediment type and supply, continental shelf slope and width, and geologic history (glaciation, volcanism, tectonic plate movement).

Categorically, major geological characteristics of coastal environments can be categorized into the following features: *rocky coasts, sandy coasts, estuaries, deltas,* and *tropical coasts.* Each coastal environment is a product of the kinds of materials present at the coast environment and the forces acting upon it. In addition, *location* plays an important role in some of the characteristics observed within a coastal environment—a topic discussed in greater detail in Section 2.4.2 under *geospatial considerations.* Each category of coastal environment identified here will be further described below with some emphasis placed on the dominating features found in each type of environment.

Rocky coasts are defined by erosive conditions brought on by high-energy impacts generated from the forces of waves, currents, and wind (Pinet 1992). As suggested in the name, rocky coasts are dominated by rock formations that can be composed of various types: sedimentary, igneous, metamorphic, etc. The kind of rocks present, particularly in the bedrock formations, will tend to dictate the overall profile of the rocky coast; erosion-resistant bedrock combined with high-energy wind and wave activity will tend to create steeper profile coastlines, whereas more easily erodable rocks (e.g., sedimentary) in low-energy environments will tend to create a more gradual coastline profile. Rocky coasts found along active tectonic margins of continental plates, particularly in areas of subduction, will tend to cause higher profile coastal environments because the geologic activity will tend to lift the area of land extruding along the coastline. Historical glacial activity may also produce

* The difference between substrate sizes can also help provide clues about the relative age of the coastline, depending on the kinds of formations found along the coast in question.

steep, rocky coasts as the glaciers cut large swaths into existing substrate revealing steep cliffs upon retreat (Pinet 1992).

There are a variety of geological features of rocky coastlines. Fjords are the results of glacial activity (mentioned above) where moving glaciers cut large swaths into existing bedrock; upon retreat, the resulting valley is eventually filled by rising seas. Many of the existing fjords today resulted from glacial advance and retreat that occurred between 2.5 million and 10,000 years ago—the Pleistocene epoch—where periods of glacier advance and retreat occurred on the Earth (Schlesinger 1997, p. 115–116).* This activity is responsible for many of the deep valley-like coastlines seen in higher latitudes. Sea cliffs are another example of a stark rise in profile between the ocean and dry uplands. The shape and appearance of sea cliffs depends on factors including bedrock material and erosion rates. Cliffs made of denser rocks, such as granite, tend to erode more evenly and thus have a uniform appearance. Softer cliff materials, such as sedimentary rock, erode more erratically due to various forces acting upon the different elevations of the cliff; the resulting appearance of softer cliffs is uneven, sometimes exposing numerous steps vertically throughout the cliff. Sea caves may be created on steeper rocky coasts exposed to intense wave forces that exacerbate erosion rates at lower elevations of the cliff. The resulting erosion helps to create bore holes, or "caves," within the cliff itself. Softer rock materials are more susceptible to the creation of sea caves but they are observed in harder substrate cliffs with strong wave activity and a long enough period of time for erosive forces to take effect.

Sandy coasts are dominated by sand, which is a substrate capable of high rates of movement and deposition over space and time. This means that sandy coastal environments are transitory and thus highly dynamic, subject to rapid and sometimes extreme changes. Where rocky coasts tend to exhibit greater levels of stability regardless of background conditions, sandy coasts are generally highly unstable and capable of rapid changes in appearance depending on the amount of energy being applied to the coastal area. Sandy coasts are typically located on passive margins characterized by low-energy wave activity, a wide continental shelf, and high offshore sediment influence. Sandy beaches tend to be areas of confluence where waves and currents converge on a particular location and deposit sand. The depositing of sand along sandy coasts produce barrier structures such as bay barriers, barrier spits, and barrier islands. Coastal barriers serve a variety of roles, including the protection of mainland areas from storm activities; marshes, tidal flats, and lagoons are often direct beneficiaries of the protections afforded by coastal barrier systems. Sandy coastlines are defined by the constant transport and deposition of sandy bottom areas with the extent of movement being defined through a mix of physical processes and general morphologic characteristics of the sea bottom (Chorley,

* There are some estimates that suggest that at the height of glacial advance during this period of Earth's history, glaciers covered approximately 30 percent of the Earth's surface.

Schumm, and Sugden 1985). A few of the features that derive from sandy coastlines include *barrier islands, barrier spits, beaches,* and *dunes.*

Barrier islands occur in areas where the physical characteristics and energy dynamics allow for the development of sandy island areas that lie parallel to the coastline. Barrier islands are dominated by features such as sand dunes, maritime forests, inlets, lagoons, back-barrier marshes, and vegetation that aids in stabilizing the less dynamic areas of the islands (Chorley, Schumm, and Sugden 1985). Without intervention, barrier islands maintain a state of dynamic equilibrium between sediment exchange, wave energy, and sea level rise. Perturbations of barrier island environments, including human manipulations, can impact the stability of these island systems by altering the dynamic equilibrium through manipulation and modification. Barrier spits are similar to barrier islands but are often attached to the landmass and generated through sediment deposition that is transported by currents and suspended by wave action. The Cape Cod National Seashore (the long "arm" off Cape Cod in Massachusetts) that runs north to south at the eastern extent of the state is an example of a barrier spit.

Beaches are sandy coastal environments that are developed through a long process of sediment deposition and movement through wind, tidal, and current action (Summerfield 1991). Beaches are often bounded on the seaside by a tidal ocean and landside by a line of vegetation. Beaches may also be defined by cliff and dune systems where the deposition and processes allow for such features to take hold. Dunes, where they exist, are large aggregations of sand that build up over time to form mounds that are generally higher than the sandy beach area immediately adjacent to the ocean. Sets of dunes can form ridges, like a mountain chain but much smaller, that lie parallel to the ocean behind the immediate sandy beach area. Dune systems form important ecological functions; they create habitat areas for unique plants and animals that thrive in dune systems, and they also provide important protection from ocean-derived storm events for areas that lie landward of the dune system. Finally, dunes offer a stabilization of beach area from the processes of erosion and also provide replacement sand for the immediate beach area that lies between the dune system and the intertidal zone.

Estuaries tend to be semi-enclosed, elongated coastal basins that receive inflows of both fresh and salt water; they often sit near the mouth of a river or watershed area that drains into a coastal basin. The mix of fresh and salt water (often referred to as "brackish" water) creates a unique environment that supports both marine and fresh-water life, depending on the diluted salinity levels occurring at different points of the estuarine environment (Schlesinger 1997, p. 279–289). Some organisms, including specialized plant species, are found only in estuary habitat as they have evolved to thrive in the brackish waters. The physical shape of estuaries often changes over time due to the high amounts of river sediment influx that settles in the estuary environment; marine sediments can also be added to estuary basins through normal tidal activity. Under certain conditions, many estuaries will become filled with sediment (depending on fluxing, rates of sediment deposition, etc.) and ultimately become watershed basin areas. Human activities can impact the rate at which estuaries are filled (through dredging, damming upriver, and other activities), and this can impact the physical changes observed in estuary systems over time. If and when estuaries become filled with sediment, the continued sediment flowing from the river is generally discharged into the ocean. When sediment supply exceeds the rate of removal by the ocean, *deltas* can form, which are a build-up of sediment at the mouth of a river emptying directly into an ocean.

Tropical coasts are coastal areas that exist within the band of latitudes that begin at the equator and move to about 30 degrees higher latitude both north and south from a beginning equatorial position. Coastal features in the tropics can be both rocky and sandy, mimicking many of the observed traits of coastal areas described above. However, warm tropical waters include prime habitat for coral, a marine organism that lives in colonies and creates extensive calcium carbonate reef systems as part of its ongoing life cycle (Summerfield, 1991). These reef systems function as a mosaic, allowing a diverse array of species to thrive. Coral reef systems can impact the overall aesthetic appearance of tropical coastal areas as the reef systems displace current activity and provide a stabilizing force for the deposition of sediment. Reefs exposed above sea level help to create landforms (via limestone formation). Submerged portions of reefs can act as barriers to mainland areas and provide some degree of storm surge protection, depending on the location and size of the reef system in relation to upland areas.

Identifying and understanding the physical characteristics of coastal areas helps paint a picture about variations in coastal features that occur in space and time. Harder coastlines made of dense materials exposed to high wave and wind energy tend to dominate northern shores. Softer coastal materials exist in equatorial and tropical areas where less wave energy is generally experienced. In between, there are numerous variations on these general themes, depending on factors that include the presence and intensity of wind, current, and tidal forces. Transport and deposition of sand under various influences provide additional factors that help shape the appearance of a coastal area. The presence of a barrier beach or island may be due to tide and current working in unison to create the particular feature observed. In turn, the created barrier offers protection to coastal inhabitants on the mainland. Understanding the relationship between major coastal features and what is helping create those features provides context when considering the impact of sea level rise on those features. Adding to the current discussion of processes and characteristics of the coastline are *geospatial* considerations; focusing on the location of the coastal feature in relation to other factors that can influence the features observed. A discussion of geospatial considerations follows.

2.4.2 Geospatial Considerations

Geospatial considerations focus on the coastline's position in relation to the ocean, for example, differentiating between a coastline facing the open ocean and one protected by an inner bay or set of barrier islands. As discussed earlier, coastlines are influenced both by the processes that impact the coast and by their actual composition; how a coastline looks is a combination of what it is made up of, the physical processes interacting with the coastline, and finally the position of the coastline in relation to those processes. This section focuses on the position of the coastline in relation to the forces that influence coastal shape and function. In addition, this section highlights the impact of large-scale coastal zone distributions, discussing potential variations in impact to sea level rise based on where the coastal zone is located across the globe.

Recall that the appearance of coastlines is often impacted by physical forces interacting with the coastal area; sandy beaches exposed to the open ocean often exist because there is low-energy wave activity at that particular area allowing the sand to accumulate along the shore. To the contrary, rocky beaches defined by larger rock and boulder sized materials along the coast often result from high-energy wave activity that transports smaller sediment away from the coast exposing the larger coastal features. There are exceptions to these rules, for example where sandy beaches accumulate in high-energy areas because of unique conditions allowing the sand to remain in an otherwise high-energy area. On a larger scale, location can impact the effects of sea level rise in coastal areas that exist in different latitudes and regions of the Earth. Coastal regions at different latitudes are subject to varying intensities of forces (gravitational, wind, current, wave), making them either more or less susceptible to the influences and impacts of sea level rise. Scaling geospatial considerations by separating larger-scale questions, such as the positioning of coastal areas at different latitudes of the Earth, from smaller-scale questions, such as the relative influence of local features on the coastal zone, provides another layer of contextual understanding when applying concepts such as shape, forces, and location of the coastal zone to further aid in understanding how these features might be impacted by sea level rise.

2.4.2.1 Small-Scale Geospatial Considerations

The relative position of a coastal area in relation to other land features and physical processes can impact the features observed. This section is meant to highlight a few of these features without going into exhaustive details on the multitude of relationships that may be found between coastal features and physical processes in a particular location. Rather than providing a full accounting of the localized relationships that can impact coastal features, the purpose of this section is to provide some awareness that the physical characteristics of coastal areas do not always fit neatly into the categories described above: Coastal areas located in high-energy environments are not always deprived of sand, and sandy areas do not always exist in low-energy environments. Importantly, a main goal of this section is to highlight that combined local features within a particular coastal zone can influence the overall impact of forces on any one of those features, sometimes mitigating

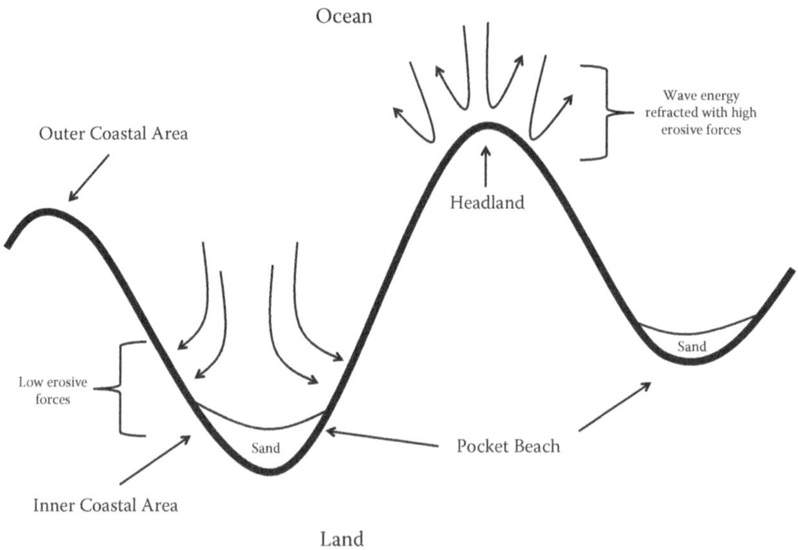

Figure 2.6 A representation of headlands and pocket beaches.

impacts and sometimes exacerbating impacts. Because sea level rise is about understanding the impacts associated with rising waters, some discussion of the dynamic relationships that can exist between multiple features is worth noting; otherwise the impacts of sea level rise might be seen to apply equally to all areas with similar conditions, which is untrue.

The position of a coastal area in relation to other nearby coastal features can impact the overall influence of each coastal feature. For example, consider the impact of wave refraction on headlands and pocket beaches; a conceptual rendering of these multiple coastal features is shown in Figure 2.6. The shape of the coast contains two parts, an outer coastal area that faces the ocean and an inner coastal area, or embayment. Wave energy is refracted around the outer headland coastal areas, contributing to erosion at the seaward portion of the coast. However, the refraction of the wave energy provides protection to the embayment areas. The embayment, in turn, collects sand over time due to the lower-intensity energy levels experienced within the embayment area. Sand deposition processes are minimally influenced by longshore currents because of the unique horseshoe configuration between the headlands and the embayment; the headlands channel current activity between the points at the outer portion of the coastline, thus protecting the inner embayment area from both wind and current. Headlands with pocket beaches that are established in high-energy areas (areas that normally do not allow for development of sandy beaches) can nevertheless provide these characteristics, depending on the relative positioning of the embayment to the outer headlands as well as the overall positioning of the landmass to prevailing currents, winds, and waves.

Spits, barrier bays, and tombolos are examples of built-up sand features along coastal areas that can alter the localized impacts depending on location. For example, a bay that is open to the ocean without any mitigating influences will be subject to the full brunt of oceanic forces, whereas a bay that is spatially closed to the open ocean or contains headland features that mitigate the effects of wave and current on the bay is more protected from direct ocean forces. However, a bay that is open to the ocean may nevertheless be protected because local conditions have allowed for the formation of a sand barrier. The sand features develop around bays or other coastal features in such as way as to provide a barrier between the effects of the open ocean and the bay itself. The barrier mitigates the impact of physical forces that may otherwise alter the appearance of the coastal area. This is particularly true where the coastal feature is facing prevailing wind, wave, and current energies; where the coastal feature is situated away from such prevailing forces the impacts are lessened, regardless of the presence of barrier features like those described here.

The relative position of a coastal feature to storm events will impact the overall effect of the storm on the coastal feature itself. For example, barrier islands tend to run parallel to larger coastal landmass features and develop as elongated strips of sandy landmass having a long length with a relatively short width (Leatherman, Zhang, and Douglas 2000). An ocean-based storm that hits a barrier island perpendicular to the island itself will tend to have a kind of impact that is different from a similar storm running parallel rather than perpendicular to the barrier island. A perpendicular impact can cause blunt erosion of certain portions of the ocean-facing island, but it can also cause *overwash,* which is sediment deposit that results from shallow water sediment being deposited on the landmass from the storm surge as it impacts the shoreline of the barrier island. A parallel storm surge is more likely to erode barrier island features without resulting in overwash.* Overwash itself is only possible where there is sufficient sediment available near the shoreline of the barrier island; depending on the location of the island, local conditions may lack sufficient substrate to allow for an overwash to occur. In these cases, the energy from the storm event may quickly erode the landmass causing a breach, or *wash,* in the structure that allows water to flow between the barrier island and directly into the inner waters adjacent to the continental coastline. Increases in sea level rise can exacerbate overwash in certain areas while also eroding the barrier island and causing breaches in barrier islands where the dynamics provide for increased energy flows with decreasing sediment transport (Leatherman, Zhang, and Douglas 2000).

Overall, small-scale variations in the conditions found along a coastal area matter. The location of the coastline in relation to other structures and processes occurring near the coastline can influence the coastline in both the appearance of the coastline itself as well as the kinds of substrate and rock formations found along

* Overwash can help maintain the integrity of the barrier island landmass by providing the landmass with nourishment by depositing ocean sediment onto the landmass.

the coastline. High-energy areas may or may not contain sandy coastal structures, depending on the coastline under observation and nearby structures that may protect that coastline from the full forces of the sea. A sense of local conditions, including the factors that establish local conditions, helps in placing the question of sea level rise impacts into context. So too does understanding the role of larger-scale geospatial considerations, specifically what roles do latitude, longitude, and landmass interactions play in affecting how coastal areas look?

2.4.2.2 Large-Scale Geospatial Considerations

The kinds of coastal environments observed vary based on latitudinal location—specifically, where the coastal feature is located on the Earth from a north to south perspective. As stated earlier in this chapter, heat energy is unequally distributed throughout the planet, causing differences in energy flow patterns. Resource distribution is also unequal, leading to rich supplies of sediment in certain regions, where other areas suffer from almost a complete lack of sediment availability. The larger-scale geospatial differences that are observed are summarized in this section.

Beaches are observed on all coastal regions of the planet, but they are composed of different materials, depending on geographic location. Sandy coasts that are shaped and maintained primarily by wave and tidal processes occupy approximately 20 percent of the global coastline (Bird 1993). These sandy coastal areas are dispersed throughout the globe, depending on the availability of sediment, physical dynamics between coastal structures, and the extent of wave and tidal energy in a particular location. A smaller proportion of global beach coastline consists of larger sediment: gravel- to boulder-sized particles. These larger sediment structured beaches tend to occur in tectonically active regions, particularly in mid- to high-latitude areas of the globe with past glacier activity. Meanwhile, coral rubble beaches and islands are common in tropical regions near the equator as the coral itself depends on clear and warmer waters that tend to be found only in tropical climates.

High-latitude coastlines are defined in many ways by cold temperatures that help to create the general morphologic characteristics observed along these regions. Sea and land ice are defining characteristics of this region and both play a role in how the coastline appears. Altering the concentration of ice through warmer temperatures can impact the morphology of high latitude coastlines in a variety of ways. Sea ice that normally buffers wave activities affecting the shore can be diminished through warmer water temperatures, allowing wave and current activities to influence coastal morphology. In addition, warmer air and water temperatures can affect the concentration of land ice, making coastal landmasses in high-latitude areas more susceptible to the forces of erosion and therefore change. Some examples of how high-latitude coastal environments can be impacted through a warming climate are outlined below.

Many coastal areas in high-latitude regions consist of unconsolidated sediments that are held together by ice. Changes in temperature can impact permafrost along the coast, changing the ice-borne consolidated landmass to unconsolidated sediment. Ground temperatures ultimately determine the presence of perennially frozen ground, which often contains large volumes of excess ice. The seasonal cycle of ground and nearshore sea water temperatures determines the depth of the seasonally active thaw layer in high latitude coastal zones. Temperature increases can limit the amount of freezing, which reduces the resiliency of the coastline to withstand the effects of erosion. In such cases, the ice that normally protects the nearshore coastline is diminished, as is the ice that consolidates the loose coastal sediments. The result can be coastal slope failure where the mitigating impact of sea ice and cold temperature sediment consolidation are removed, exposing the coastline to erosive forces where it otherwise would be buttressed by the ice against such forces (Dallimore, Wolfe, and Soloman 1996).

Sea ice has the capacity to diminish wave activity by providing a buffer between the open ocean and the coastal area, mitigating the impacts of ocean-derived storms on the coastline by absorbing the wind and wave energy before it hits the coastline. Sea ice can also supply sediments to the coastline; as the sea ice develops it drags along coastal seabed areas bringing sediment to the coast. When warmer temperatures allow the ice to retreat, the sediments are dislodged, providing a source of sediment build-up along these coastal regions (Reimintz, Barnes, and Harper 1990). Warmer temperatures can disrupt this process; less sea ice is available as protection as warmer temperatures melt the icepack earlier each year, exposing the coastal zone to longer periods of wind, tide, and current interactions. In addition, a thinner icepack provides less sediment accumulation, meaning that less sediment is available to coastal areas. The result is an increased capacity for erosion along coastal regions in higher latitudes with less sediment deposition potential through sea ice transport.

Studies have been done to calculate the effects of warming trends on Arctic shorelines as a means of understanding the long-term impacts these warming trends might have on high-latitude coastal environments that have developed around icepack conditions. For example, coastal erosion rates of high-latitude coastlines have been correlated with the impact of open water (iceless water) exposure of these coastlines to show a rate of erosion that increases with the amount of time these coastlines remain unprotected by icepack; essentially, the more time the coastline is exposed to water, the greater the erosion of that coastline (Forbes and Taylor 1994). This is particularly true in certain regions of high-latitude coastline such as the Canadian Arctic Archipelago region, where many fine-grained (mud and sand) shorelines and deltas now experience almost zero wave energy due to icepack buffering; an increase in open water exposure will likely lead to the potential for exacerbated shoreline retreat in these areas because of the particularly susceptible unconsolidated and fine-grained makeup of the shoreline substrate (Forbes and Taylor 1994).

Rapid coastal retreat is already being observed in high-latitude coastal areas defined in large part by icepack and coastal landmass that is held together by permafrost. Coastlines in the Beaufort Sea in northwestern Canada (Dallimore, Wolfe, and Soloman 1996) as well as Arctic coastlines of the United States and Russia present examples of high-latitude coastal areas where prolonged coastal retreat has been observed and correlated to warming trends that include reduced sea ice and permafrost along coastal features. The implications suggest that warming trends have a particular impact on the overall morphology of high-latitude coastal features, presenting special considerations for adaptive strategies geared toward sea level rise. For example, coastal communities in these regions often exist in land–sea interfaces defined by ice; if the ice erodes, then the viability of these communities is put at risk due to the cumulative effects that warming has on high-latitude coastal features (Forbes and Taylor 1994).

Tropical coastlines are dominated by reef systems that are created by coral organisms and exist in a variety of forms, including: fringing reefs, barrier reefs, and atolls. The communities that develop around coral reefs are species diverse, providing a rich source of biodiversity within reef systems that are utilized for a variety of human purposes, including as a source of food for coastal communities (Wilkinson et al. 1999). Coral reef systems also serve as a viable landmass resource in tropical regions of the ocean, an area that is nutrient and resource limited. In addition, coral reef systems provide important ecosystem regulating services, including coastal protection from storms (Moberg and Folke 1999).

Increased carbon dioxide emissions into the atmosphere can impact coral reef communities, the most direct impact being the slow acidification of the ocean as increased atmospheric carbon dioxide is taken up by the oceans through air–sea exchange patterns (Schlesinger 1997). The process of carbon uptake by the ocean mixes carbon dioxide with water to create carbonic acid in a carbonate cycle that includes a multistep chemical reaction process. The increase in carbonic acid in the shallow depths of the ocean increases the reactivity of the sea water, particularly in relation to the carbonate that helps to form the structure of the coral shells. This change in chemistry of shallow coastal waters can impact the ability of the coral to survive, leading to "bleaching" effects where coral reef systems appear white as the organisms begin to die off. The end result can be a diminished capacity of coral organisms to continue the development of coral reef systems and thus decrease the availability of coral itself as a limited resource for the coastal features observed in tropical regions of the world's oceans.

The location of coastal features relative to the Earth's latitude impacts the actual feature observed. High-latitude coastal areas tend to have features dependent on freezing temperatures that allow for conditions of sea ice and permafrost. Remove the sea ice and permafrost from these regions and the coastal features change. The same is true of tropical regions at low latitudes. These coastal areas are dominated by the presence of coral, which serves as a critical resource in coastal community and structure development. Remove the coral from tropical waters and the coastal area changes. Combining an understanding of latitudinal variations in coastal features (large-scale geospatial considerations) with more detailed variations of coastal areas in relation to other variables (proximity to other landmasses, relationship to tidal and wind influences, current considerations, etc.) begins to paint a more detailed picture of the potential impact of sea level rise and the associated phenomena (particularly climate change) that accompanies sea level rise. One additional component to consider is the role of human development in helping to shape coastal land use features; this component is discussed next.

2.4.3 Development Considerations

Human development along coastal regions has impacted the features seen along coastal areas. Some of the mechanisms impacting coastal features have already been mentioned above. For example, jetties and groins are human-created structures that run perpendicular to coastal areas into the sea. These structures often cut off the natural flow of longshore currents, building up sediment along one side of the structure while depriving sediment to coastal areas on the other side of the structure. The effect of such structures alters the coastal land features, which can have consequences for the overall dynamic interactions of the area. Blocking sediment transport along a shoreline limits the amount of sediment available down-current from the blocking structure, essentially depriving a portion of the coastline of a source of nourishment to sustain and build coastal landmass. Over a period of time, this lack of sediment nourishment can lead to higher erosion rates along a portion of a coastal region, which can be exacerbated by sea level rise as these areas lack sediment deposition to mitigate the impacts of the rising tide and thereby allowing sea level migration landward to areas that are located at or near sea level.

Armoring against the sea has also been mentioned as a human-based activity that is normally found in developed coastal areas. Deciding to develop and establish communities along a coastline creates commitments both today and certainly in a future of sea level rise. Establishing a community means creating physical structures that are bounded to the land. In addition, social institutions are developed as governments are formed and the community is officially incorporated into the region. Often, public services are created in these areas that include the development of infrastructure, including piping for sewage and water dispersal, utility provisions, roads, and other structures that are heavily site dependent, meaning the structures often remain in the physical location in which they are placed. Most of these structures (buildings, infrastructure, etc.) are connected to the land and contain substantial up-front costs, representing a front-ended commitment by those living in the region to remain in the area for a long period of time. Because of these commitments, mechanisms are often developed to protect the region against potential influences from nature; armoring is one method developed to prevent the effects of storms and sea level fluctuations from impacting the coastal community.

Armoring impacts coastal land use features by displacing the normal transition zones often seen in coastal zones (see Figure 2.5) with a stark demarcation between land and sea (at least where sea level has risen to meet the armoring structure). This is overtly seen in instances where seawalls are created; the ocean often meets the sea wall during maximum tidal range or during its entire tidal range. The wall itself provides a hard line upon which the coastal area is defined, at least as long as the sea does not breach the upper height of the wall. As sea level increases over time, it may be that human armoring structures are breached, either through storm events such as Hurricane Katrina or through rises in average sea level. The effect of the breach will likely cause a substantial change in coastal land use features as the wall prevented the normal encroachment of the sea and once this unnatural holdback of the ocean has been released, a stark punctuation to a new equilibrium state will be achieved as the sea moves quickly landward, uninhibited by the armoring structure.

Dredging and filling are other examples of development that impact the appearance of the coastline. Dredging is the purposeful removal of sediment material from the water, usually done to maintain a certain depth in the coastal area or to ensure the continued flow of water along a particular area of the coastline. Recall the formation of deltas and longshore currents, two methods by which sediment is carried by a water body from one location to another where it is deposited. Deltas are formed when rivers carry sediment to the area where the river meets the ocean (the "mouth" of the river); when the quantity of sediment being deposited at the intersection of the river and ocean exceeds the amount of sediment being removed by the ocean waves and currents, the sediment accumulates, forming a delta, or landmass, around the intersection. Longshore currents often pick up sediment from one location and bring it to another where physical forces and conditions allow the sediment to deposit. Over time, the sediment can build up to a point where it exceeds the height of the sea, forming a variety of structures.

Dredging is the process by which humans, for a variety of purposes, actively remove the sediment built up in areas. Dredging often occurs around bays and harbors to ensure a minimum depth of water for commercial and recreational shipping purposes, allowing the ships to gain access to nearshore areas. Filling is where humans will take sediment from one location and use it to purposefully add sediment to a different location. Sometimes this is done on land to bolster the landmass near a coastal area, as is often done to maintain sand on a beach area by enhancing the existing sandy beach or creating dunes to protect the area against the forces of erosion. Other times, it is done to "nourish" landmass areas that are low lying, essentially building up landmass to protect against rising tides. In certain cases, filling is done to create a new landmass substrate for development purposes, as when wetland areas are filled so that the wetland becomes dry land and can be used to place buildings or other structures on the newly created dry land.

The manipulation of coastal land use features by humans is reaching new heights in certain areas of the world. Where land is scarce, or new ocean communities are desired, human ingenuity employs technological innovations to provide coastal feature development that matches aesthetic interests. In the Middle East, new landmasses are being built to allow for the development of residential and recreational structures along bays and ocean areas. This is a different kind of coastal land alteration; there is no alteration of existing coastline but rather the creation of entirely new coastal areas, including peninsulas and islands. The extent of these structures and their potential impacts on coastal ecology and processes is not fully known, but the structures are certainly presenting an interesting dilemma as one considers a future where sea level rise is occurring. Purposefully committing humans to a sensitive area of the coast when the impacts of sea level rise are not fully known presents public policy considerations that include an upper limit on government's capacity to ensure the safety and well-being of these communities over the medium to long term. One needs to question whether the people investing in these communities are well aware of the potential risks associated with their investment? One also needs to consider whether government is engaging in their full responsibility to assess these risks before committing infrastructure resources for the development and servicing of these structures.

There are many other kinds of human development interactions with the coastline that impact the way in which a coastline looks and functions. A number of these particular human structures are listed in the policy section of this text (Chapter 3). However, the examples provided here show that human development not only impacts the way in which a coastline looks physically, but also the way in which the coastline functions from a bio-geophysical standpoint. Ecosystems are often disrupted through human actions of armoring against incoming seas; wetlands and intertidal zones are diminished through this process, affecting biodiversity that would normally develop without the structure in place. In addition, the acts of armoring, dredging, and filling all work to provide humans with a sense of security along coastal areas that only reinforces the commitment to live within those coastal areas. This commitment is explored in greater detail in Chapter 3, but it is certainly important to begin thinking about the sense of creating expectations of commitment in a zone that is highly dynamic and changing. If sea level rise continues to its maximum potential as mentioned earlier in this section, then we as a society likely need to rethink the logic behind our purposeful actions that reinforce an expectation of commitment to developing and maintaining communities in coastal areas. Choices beyond full commitment to these regions will be explored in more detail as we move outside the science and observations of coastal areas to the law and policy discussions that follow in the remainder of this text.

2.5 Conclusion

This chapter explored the scientific principles behind sea level rise. Specifically, the relationship between physical and chemical interactions was the basis for explaining three main things: (1) current observations related to sea level rise, (2) future predictions of the extent of sea level rise, and (3) major coastal land features connected with sea level rise. This breakdown is meant to provide the reader with a baseline understanding of how chemical and physical interactions are connecting to cause the sea level rise observed today with future predictions of the extent of that sea level rise, and also how the features of a coastline may either mitigate or exacerbate the effect of sea level rise. With this foundational understanding, the policy and legal implications can be placed into a context so that options surrounding how to deal with observed and future predictions of sea level rise can be made based on the best evidence available to protect the interests of human beings both today and in the future.

Fundamental concepts discussed in this chapter include understanding the interactions of energy flows within the Earth as a functioning system. There are critical points to remember from this section. First, the Earth can be seen as a system itself with a number of subcomponent systems. Viewing the Earth as a system with subsets of systems helps to place energy flow dynamics between system components into a perspective that allows for an understanding of sea level rise based on heat energy distributions. In summary, heat energy from the sun is being retained in the Earth system due to increased amounts of carbon in the Earth's atmosphere. The additional trapped heat is causing changes in the water cycle, specifically water trapped as ice to melt and also thermal expansion of water molecules; both of these products of trapped heat are adding to sea level rise. This is what is seen today through current observations linking carbon increases in the atmosphere with melting icecaps and glaciers, as well as the thermal expansion of ocean waters.

The second critical point is that feedback mechanisms within the system will reinforce further heating without any changes made to the heat energy transfer from the atmosphere to the water cycle. The melting of glaciers and icecaps is releasing ice and snow cover from the land, exposing darker-colored space underneath the ice. The darker space retains heat where the ice and snow once reflected heat. Greater retention of solar energy reinforces the heating of the Earth system, which reinforces the melting of more ice and thus reinforces sea level rise. These feedback indicators suggest heating begets heating and this can mean—absent contrary feedback mechanisms—that more sea level rise is likely to occur in the future. The fact that carbon forcing—removing stored sources of carbon from the ground (lithosphere) and moving the carbon into the atmosphere—seems to support the presumption that a warming trend will continue. If so, we can legitimately expect sea level rise to continue. Thus, the impact of sea level rise may be legitimately considered an ongoing concern for policy makers and coastal managers.

The links between observations today and future predictions help provide a sense of the extent of sea level rise over time. A conservative estimate using current rates of sea level rise continuing over the next several decades creates a baseline from which policy planning may occur. The worst-case scenarios of total glacial melting of large ice structures, such as the Greenland Ice Sheet or large ice shelf areas in the Antarctic, provides an outer estimate of how much sea level can be expected to rise under these worst-case scenarios. Planning can take these different scenarios into consideration and use them to determine proactive courses of action that balance the needs of society today with the likelihood of impacts to coastal areas to society over time. The science of understanding sea level rise helps place planning into a context where risks can be weighted and decisions made based on rational estimates.

Understanding the morphology of coastal features—the way coastal features look—adds another element to assessing the potential impacts of sea level rise over time. The makeup of coastal features, where they are located, and human alterations of coastlines are all factors that influence the overall impact of sea level rise today and in the future. Coastal areas made up of denser materials and exhibiting some height above sea level will, all other factors being equal, tend to weather sea level rise better than coastal areas dominated by softer substrate features that lie at or near sea level. In addition, the relative location of coastlines matters; coastal areas located in higher latitudes often include ice and freezing temperatures as part of the chemistry that helps to define coastal features in these regions; melting and thawing coastal permafrost can allow for changes that increase erosion and thus exacerbate the effects of a warming planet in high latitude regions. Tropical regions may also be influenced because of the sensitivity of coral reefs to chemistry changes in the acidity of the water brought on by increased carbon concentrations. Coral is a primary structural source for tropical coastal areas and thus impacts on the health of corals can influence the physical features observed in tropical coastlines. Finally, human development along the coast can provide a number of soft and hard structural elements that alter both the existing appearance of the coastline as well as the capacity of the coastline to be altered by changes that accompany sea level rise. To truly understand the impacts of sea level rise locally, one must understand the unique features exhibited at the coastal area in question.

The background provided in this chapter now sets the stage for a more in-depth discussion of *policy options* and *legal considerations*. The chapter on policy options (Chapter 3) focuses on approaches to sea level rise that surround adaptation policies, essentially how to adapt to sea level rise at the coast. After adaptation strategies are discussed in general, a discussion of different considerations *between* adaptation strategies will follow to help place the strategies discussed into a larger context that considers policy choices from a set of economic, social, and ecological perspectives. Following the policy chapter, an in-depth exploration of legal considerations related to managing sea level rise in the coastal zone occurs in Chapter 4. The perspective taken here is one of constitutional limitations that limit policy directions. The goal is to provide managers and policy makers with greater insights into some of the legal hurdles that exist in proactive coastal zone management of sea level rise in light of United States legal precedent. Recommendations are made that help decision makers consider how to manage sea level rise in a way that mitigates the legal issues presented.

As this discussion moves forward into a policy/legal dynamic, it is critical to bring along the concepts discussed in this chapter because these science concepts provide the outer set of principles that define and ground the policy and legal debates in a framework of how the Earth actually operates. The goal is to develop policy directions that internalize natural conditions so those directions are in harmony with the principles that are actually driving sea level rise. In doing so, policy development can be harmonized with ecosystem principles, ensuring that the adaptation approaches undertaken are the kinds best suited to maximize value for humans at the coastline both today and tomorrow.

References

Adger, W., N. Vincent, and K. Vincent. 2005. Uncertainty in adaptive capacity. *Geoscience*, 337: 399–410.

Bamber, J., R. Riva, B. Vermeersen, and A. LeBrocq. 2009. Reassessment of the potential sea-level rise from a collapse of the West Antarctic ice sheet. *Science*, 324: 901.

Barnett, J., and J. Campbell. 2010. *Climate Change and Small Island States: Power, Knowledge and the South Pacific*. London: Earthscan.

Bertaianffy, L. 1968. *General Systems Theory: Foundations, Development, Applications*. New York: George Braziller.

Beever, L. 2009. Climate ready estuaries: Vulnerabilities and adaptations. *15th Annual Public Interest Environment Conference, Levin College of Law*. University of Florida, Gainesville, February 29, 2009.

Bird, E. 1993. *Submerging Coasts: The Effects of a Rising Sea Level on Coastal Environments*. Chichester, UK: John Wiley & Sons.

Brooks, N., N. Adger, and M. Kelly. 2005. The determinants of vulnerability and adaptive capacity at the national level and the implications for adaptation. *Global Environmental Change*, 15: 151.

Bryant, E. 1997. *Climate Process and Change*. Cambridge (UK) and New York: Cambridge University Press.

Chorley, R., S. Schumm, and D. Sugden. 1985. *Geomorphology*. London: Methuen.

Costanza, R., R. d'Arge, R. de Groot, et al. 1997. The value of the world's ecosystems services and natural capital. *Nature*, 387: 253.

Dallimore, S., S. Wolfe, and S. Solomon. 1996. Influence of ground ice and permafrost on coastal evolution, Richards Island, Beaufort Sea Coast, NWT. *Canadian Journal of Earth Sciences*, 33: 664.

Dasgupta, S., B. Laplante, C. Meisner, et al. 2007. The Impact of Sea Level Rise on Developing Countries: A Comparative Analysis. World Bank Policy Research Working Paper No. 4136, The World Bank, Washington D.C.

Fatowsky, D., and P. Sheehan. 2005. The extinction of the dinosaurs in North America. *GSA Today*, 15(3): 4.

Finan, T., C. West, T. McGuire, and D. Austin. 2002. Processes of adaptation to climate variability: A case study from the US Southwest. *Climate Research*, 21: 299.

Forbes, D., and R. Taylor. 1994. Ice in the shore zone and the geomorphology of cold coasts. *Progress in Physical Geography*, 18: 59.

Gehrels, W., B. Hayward, R. Newnham, and K. Southall. 2008. A 20th century acceleration in sea-level rise in New Zealand. *Geophysical Research Letters*, 35: L02717.

Hemming, D., J. Iowe, M. Biginton, et al. 2007. Impacts of Mean Sea Level Rise Based on Current State-of-the-Art Modeling. Hadley Centre for Climate Prediction and Research, Exeter, UK.

Intergovernmental Panel on Climate Change. 2001. IPCC Third Assessment Report: Climate Change 2001 (AR3). http://www.ipcc.ch/publications_and_data/publications_and_data_reports.shtml (accessed August 8, 2012).

Intergovernmental Panel on Climate Change. 2007. IPCC Fourth Assessment Report: Climate Change 2007 (AR4). http://www.ipcc.ch/publications_and_data/publications_and_data_reports.shtml (accessed August 7, 2012).

Jerejeva, S., A. Grinsted, J. Moore, and S. Holgate. 2006. Nonlinear trends and multiyear cycles in sea level records. *Journal of Geophysical Research*, 111: C09012.

Kalo, J., R. Hildreth, A. Rieser, et al. 2007. *Coastal and Ocean Law: Cases and Materials*. St. Paul, MN: Thompson/West.

Keeling, C. 1960. The concentration and isotopic abundances of carbon dioxide in the atmosphere. *Tellus*, 12: 200.

Kelly, S. 2000. The Utilization of Seawalls in Response to Shoreline Erosion: Consequences, Socio-Economic, Political and Legal Forces, and Alternatives. Donald Bren School of Environmental Science and Management, University of California, Santa Barbara. http://ocpc.msi.ucsb.edu/pdfs/Seawall.pdf (accessed August 8, 2012).

Lambeck, K., M. Anzidei, F. Antonioli, et al. 2004. Sea level in Roman time in the Central Mediterranean and implications for recent change. *Earth and Planetary Science Letters*, 224(3-4): 563.

Leatherman, S., K. Zhang, and B. Douglas. 2000. Sea level rise shown to drive coastal erosion. *Eos*, 81(6): 55.

Margulis, L., and D. Sagan. 1995. *What is Life?* Berkeley, CA: University of California Press.

McGuire, C. 2012. *Environmental Decision-Making in Context: A Toolbox*. Boca Raton, FL: CRC Press.

Meyerson, L. 1999. Ecosystem-Level Effects of Invasive Species: A Phragmites Case Study in Two Freshwater Tidal Marsh Ecosystems on the Connecticut River. Yale University School of Forestry and Environmental Studies. http://nrs.uri.edu/labs/invasive/PdfReprints/Full_Thesis_PDF.pdf (accessed August 8, 2012).

Millennium Ecosystem Assessment. 2003. Ecosystems and Human Well-Being: A Framework for Assessment. http://www.maweb.org/en/Framework.aspx (accessed August 8, 2012).

Moberg, F., and C. Folke. 1999. Ecological goods and services of coral reef ecosystems. *Ecological Economics*, 29: 215.

Molles, M. 2008. *Ecology: Concepts and Applications, 4th ed.* New York: McGraw-Hill.

Nicholls, R., and J. Hoozemans. 2005. Global vulnerability analysis. In *Encyclopedia of Coastal Science*, edited by M. Schwartz, 486-491. Dordrecht: Kluwer Academic Publishers.

Nicholls, R., and R. Tol. 2006. Impacts and responses to sea-level rise: A global analysis of the SRES scenarios over the twenty-first century. *Philosophical Transactions of the Royal Society*, 364: 1073.

Oliver-Smith, A. 2004. Theorizing vulnerability in a globalized world: A political ecological perspective. In *Mapping Vulnerability: Disasters, Development and People*, edited by G. Bankoff, G. Frerks, and D. Hilhorst, p. 10–24. London: Earthscan.

Pelling, M., and J. Uitto. 2001. Small island developing states: Natural disaster vulnerability and global change. *Environmental Hazards*, 3: 49.

Philander, G. 2008. *Encyclopedia of Global Warming and Climate Change.* Thousand Oaks, CA: Sage Publications.

Pinet, P. 1992. *Oceanography: An Introduction to the Planet Oceanus.* St. Paul, MN: West Publishing Co.

Randolph, J. 2004. *Environmental Land Use Planning and Management.* Washington, D.C.: Island Press.

Reimnitz, E., P. Barnes, and J. Harper. 1990, A review of beach nourishment from ice transport of shoreface materials, Beaufort Sea, Alaska. *Journal of Coastal Research*, 6: 439.

Riggs, S., and D. Ames. 2003. Drowning of the North Carolina Coast: Sea-Level Rise and Estuarine Dynamics. North Carolina Sea Grant, Publication No. UNC-SG-03-04, 152 pp.

Ringot, E., and P. Kanagrantnam. 2006. Changes in the velocity structure of the Greenland Ice Sheet. *Science*, 311: 986.

Schlesinger, W. 1997. *Biogeochemistry: An Analysis of Global Change, 2nd ed.* San Diego, CA: Academic Press.

Summerfield, M. 1991. *Global Geomorphology: An Introduction to the Study of Landforms.* New York: Longman.

Titus, J. 1998. Rising seas, coastal erosion, and the takings clause: How to save wetlands and beaches without hurting property owners. *Maryland Law Review*, 57: 1279.

Tol, R., M. Bohn, T. Downing, et al. 2006. Adaptation to five meters of sea-level rise. *Journal of Risk Research*, 9: 467.

U.S. Census. 2011. NOAA's State of the Coast. http://stateofthecoast.noaa.gov/population/welcome.html (accessed August 8, 2012).

U.S. Global Change Research Program. 2009. Global Climate Change Impacts in the United States (18). http://www.globalchange.gov/what-we-do/assessment/previous-assessments/global-climate-change-impacts-in-the-us-2009 (accessed August 7, 2012).

Wada, Y., L. van Beek, F. Welland, et al. 2012. Past and future contribution of global groundwater depletion to sea-level rise. *Geophysical Research Letters*, 39: 6 pp.

Wilkinson, C., O. Linden, H. Cesar, et al. 1999. Ecological and socioeconomic impacts of 1998 coral mortality in the Indian Ocean: An ENSO impact and a warning of future change? *Ambio*, 28(2): 190.

Chapter 3

Policy Options

3.1 Introduction

Chapter 2 provided an overview of the science behind sea level rise. Included in that overview was a discussion of current observations and future projections of sea level rise. What is known from that overview is that sea level is rising and there is a strong possibility that sea levels will continue rising for decades to come. What is not precisely known is just how far seas will actually rise. Will they rise by a foot? By 10 feet? Knowing that sea level is rising, but not knowing the extent of that rise creates uncertainty about the phenomenon; the extent of uncertainty is mitigated in part by a deep understanding of the science that drives sea level rise, but even a deep understanding of the science does not remove the uncertainty that accompanies sea level rise discussions.

Uncertainty about sea level rise is an important factor in this chapter on policy options. Policy is fundamentally a discussion about what is being done to solve a problem or address an issue, and *public policy* is about the choices that government makes to solve a problem or address an issue. This chapter focuses on policy options associated with sea level rise. Essentially, what are the available options to deal with the seas as they encroach upon the shores? And where more than one option exists, what is the preferred option? There is little to debate about the uncertainty of existing sea level rise because it can be currently observed (see Chapter 2). Because current sea level rise provides little to no uncertainty in terms of its existence, policy choices can be made that account for what is observed in terms of sea level rise today. However, *future* sea level rise is full of uncertainty and it is that uncertainty that makes policy decisions more difficult. For example, does it make sense to place restrictions on the current use of coastal land to prevent the *possibility* of harm associated with sea level rise in the future? What if those current restrictions place substantial economic costs on society today with no assurance those costs will achieve benefits for members of society today or in the future? In essence, how practical is it to make choices about interactions in the coastal zone today when there is imperfect information about what will happen to our coastal zone in the future? These questions are the foundation upon which policy discussions related to current and future sea level rise begin.

This chapter discusses several categorical options available to coastal managers in adapting to sea level rise. It begins the discussion by identifying the larger thematic issues that present themselves when entertaining policy options that require making choices in the face of uncertainty. It makes sense to prepare coastal structures today for sea level rise that will result in the inundation of those structures in 10 years. However, it may be a misuse of capital to create unnecessary resiliency in these structures—at additional cost—if seas do not rise sufficiently to impact these structures. Similarly, it makes sense to prevent development along a coastline "buffer zone" to ensure there is sufficient open space to allow for wetland migration landward as sea level rises. However, if sea level does not rise high enough to justify the land left untouched, then the unused land may be seen as a lost opportunity for coastal development, including all the economic benefits that derive from such development.* The uncertainty that accompanies future sea level rise impacts policy directions; these impacts are raised in the thematic issues discussion where policy questions related to sea level rise are framed.

After considering some of the larger policy questions associated with sea level rise, the chapter moves on to adaptation approaches. The approaches discussed are categorically separated into two main choices: stay or retreat. *Staying* means remaining committed physically and resource-wise to the coast. Under this philosophy, two main categorical choices are made: (1) protect against the incoming sea, or (2) adapt to the incoming sea. In addition, some hybrid choices that accompany staying are discussed, where a level of protection and adaptation are combined, depending on the extent of sea level rise encountered. The alternative to staying—*retreating*—means moving away from the water as the sea begins to move landward. Retreat is not meant to signify a strong single movement away from the coast, but rather a continuum where a mix of adaptive measures to mitigate impacts in the short term are combined with a planned retreat as sea level rises to some maximum point along the coastline in question.

The final subject discussed in this chapter offers the reader a consideration between the adaptation approaches discussed earlier in the chapter. Each adaptation choice carries economic, social, and ecological considerations that impact the decision-making process. For example, the choice to stay and armor against the tide might result in a commitment of resources that cannot be mitigated should sea level rise too high and the area needs to be abandoned; the economic and social

* These benefits can include a substantial property tax base from higher-value coastal land that is developed and privately owned.

costs can be substantial under such circumstances. Meanwhile, the choice to armor can impact ecological resources by removing essential habitat for coastal features such as wetlands to migrate landward. The costs associated with the loss of wetland habitat and similar ecological values can be substantial. Alternatively, the choice to retreat from the area can lead to immediate costs economically and socially as communities purposefully retard natural growth patterns and preferences through proactive policies that prohibit the development or armoring of land to protect against a future sea level rise that may or may not occur. Understanding the tradeoffs that are inherent in the different adaptation choices available is an important part of considering the overall impact of policy choices related to sea level rise.

Policy choices surrounding sea level rise are complicated and difficult: complicated because they include choices that are based on evolving and new ways of human interactions at the coastline, and difficult because they include making decisions in the face of uncertainty. *Adaptation* policy choices begin from a presumption that sea level rise is occurring and will continue to occur in the future. This presumption alone makes the kinds of policy choices encountered even more difficult because of the uncertainty associated with the future extent of sea level rise. One way these choices can be simplified is to identify policy options that create benefits *regardless* of the extent of sea level rise encountered, and distinguish these options from policy choices that are beneficial only under circumstances where future sea level rise is moderate to substantial. This is one way to mitigate the uncertainty associated with adaptation policies related to sea level rise, and it will be one of the frameworks established in this chapter used to compare between policy options in each category. If, for example, a policy option could be identified that may be utilized under any circumstance and allows the greatest degrees of individual freedom in adapting to sea level rise, then this choice may be superior to other choices because of its general applicability. However, the term *superior* must be carefully defined within this context because a policy option that might achieve a large direct economic benefit at little cost may nevertheless contain high ecological costs that, on balance, may prove the policy option to be inferior, depending on the values being considered.

To gain clarity over the different subjects discussed in this chapter, topics are segregated into categorically distinctive areas—for example, *stay* versus *retreat*—as a means of creating a dialectic among policy choices; one can move in a policy direction that allows a community or individual to stay at the coast, or one can move in a policy direction that ultimately involves retreating from the coast. In reality, these two categories represent ends of a spectrum where there are (or may be) infinite policy approaches between fully staying and fully retreating from sea level rise. The actual choice may contain elements of both staying and retreating over a larger policy framework to adapt to sea level rise. For example, beach nourishment projects may be used to mitigate the near-term impacts of rising seas, allowing the water to be held back for a longer period of time than if nothing were done. In this way, coastal property owners have more time to *stay* along the coastline. This short-term policy may be combined with a longer-term policy where insurance risks are incrementally passed from a shared private/government distribution of risk to a wholly private distribution, thus allowing the private landowner to internalize the increasing risks associated with rising tides as the premiums for insuring private property increase to reflect changes in coastal dynamics. The risk shifting of insurance premiums may be coupled with regulations that prohibit the construction of seawalls and other hard-armoring devices to allow associated coastal features (wetlands, vegetation, etc.) to migrate landward as the sea rises. Collectively, these policies can work in concert to provide a mix of features that allows people to stay near the coast for longer periods of time while also providing incentives to retreat from the coast when sea levels have risen to a point that justifies retreat.

The simple example above shows how a mix of policy choices working in concert over time can help mitigate the impacts of sea level rise through adaptation strategies that allow for some continued use of the coastal resource while also preparing for advancing tides. The materials provided in this chapter explore these different policy options and instruments in greater detail. However, it is important to understand that none of these policy choices operates in isolation. Rather, each category of policy options should be viewed as a tool made available to those involved in coastal management planning and decision making; sometimes one policy tool may be superior to all other choices for a particular situation, and other times a variety of policy tools may be used in combination to achieve a particular goal. The purpose of this chapter is to provide an overview of the available tools within a context of uncertainty. The manner in which these tools are applied will depend on the factors involved in real-life situations to allow for practical policy solutions to the problem that is sea level rise.

3.2 Framing the Policy Questions Related to Sea Level Rise

The policy questions related to sea level rise are influenced by the *uncertainty* of how much the seas will rise in the future, calling into question the ultimate impact of sea level rise on a particular area. Policy questions are also influenced by the factors used to judge policy directions; this chapter categorizes those factors into three main areas: (1) *economic,* (2) *social,* and (3) *ecological* (these three factors are addressed in greater detail later in this chapter). The goal of this first section is to provide a general framework for understanding the relationship between policy goals in general and the particular issues raised when considering sea level rise adaptation strategies in the context of uncertainty, including the impacts decisions today may have on the economic, social, and ecological welfare of coastal regions.

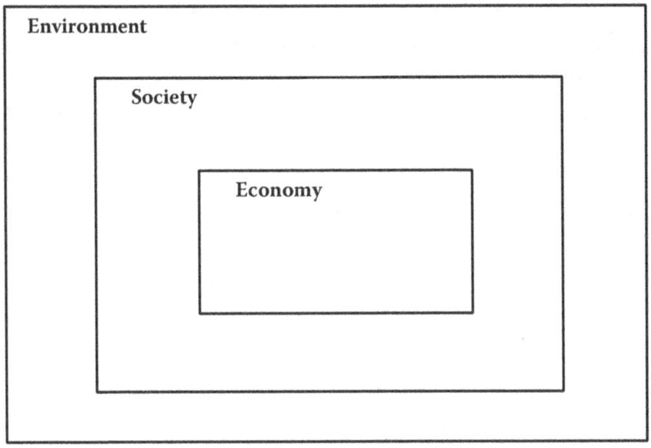

Figure 3.1 Relationship between environment, society, and economy.

To begin, a reference point must be established that provides a conceptual framework for judging policy options related to sea level rise. Figure 3.1 provides a conceptual framework that creates a relationship between three important factors: environment, society, and economy.

As depicted in Figure 3.1, society and the economic activity generated from social interactions are within the larger box of the environment. This relationship is meant to suggest that the environment provides the outer limits by which human society functions; *the presumption here is that society is limited by the environment in which the society forms and develops*. Because the economy "box" sits within society in this figure, it suggests that the economy is also limited by the environment as well as society. Taken holistically, the figure represents an important presumption about the interactions between humans and the environment, suggesting that the environment provides the outer limit by which humans function. If the environment is degraded, the degradation can impact the capacity of human society to include economic activity.*

The presumption stated in Figure 3.1 provides an important foundation for policy decisions in general. In short, it suggests that policy choices (what we choose to do) are constrained by environmental limits. If this is true, then the environment becomes an important primary consideration in making policy choices; choices that degrade background environmental conditions have the capacity to limit future choices, while policy choices that preserve background environmental conditions sustain, and potentially expand, policy options in the future. If this premise of environmental constraints is accepted, then a policy preference is established

* This presumption fits within the discussion in Chapter 2 on the Earth as a system of processes that interact with one another and the importance of understanding how perturbations to those interactions can affect the way in which the Earth operates.

where a set of choices presented can be rank ordered by the impact those choices have on background environmental conditions; all things being equal, choices that preserve background environmental conditions are superior to choices that degrade background environmental conditions. Similarly, choices that provide less harm to background environmental conditions are superior to choices that provide greater harm to background environmental conditions.

The initial presumption between the environment, society, and the economy identified above is used to ground policy making in relation to sea level rise adaptation and is a key starting point to making decisions between several options presented. As shown later in this chapter, choices to stay and armor the coastline present a number of benefits, particularly where existing infrastructure and coastal communities are concerned. However, the option of armoring often carries environmental consequences that can impact the capacity of the environment in that region. If the choices between armoring and other policy responses are weighed against the presumption of environmental conditions as critical to overall social and economic well-being, then the superior policy choice may be more difficult to determine than if the focus is simply on the immediate safety of the existing coastal community. Often, policy choices are not solely about making decisions that focus on the immediate benefits of those living today, but they also need to include some consideration about the impact of choices today over a longer period of time. Considering the wider impact of policy choices over a longer period of time is often referred to as the *sustainability criterion* (Bowersox 2002).*

Once the initial presumptions are established, the specific policy goals must be understood. In this text, the focus is on *adaptation* strategies related to sea level rise. Taking a moment to reflect on the sequencing of adaptation in an overall planning strategy, the focus on adaptation can be placed into a larger context to understand what constraints, if any, exist on the framework from which adaptation goals are being established. Consider that adaptation is but one response to sea level rise. There are also other responses that may include *prevention* and *mitigation* of sea level rise, both of which place the role of policy decisions into a different context. Understanding how adaptation policy development and implementation differs from prevention or mitigation measures is important to establish a baseline from which policy approaches are being created in the adaptation context. Similarly, understanding the role prevention and/or mitigation strategies might play in aiding

* The *sustainability criterion* is used as a way of "measuring" the policy choice against criteria that ensures the decisions provide for humans today without diminishing the capacity of the environment to provide for humans in future generations. Inherent in this statement is the suggestion that the environment is the background condition against which all other human decisions must be judged. Where the environment is being diminished in its capacity to provide for future generations, the sustainability criterion is not being met. Where the environment is being sustained and its capacity to provide for future generations is maintained, the sustainability criterion is being met. The sustainability criterion will become an important theme when the ecological impacts of adaptation choices are examined later in this chapter.

adaptation policy development can be useful when thinking holistically about responding to sea level rise.

Prevention of sea level rise requires some understanding of the *causes* of sea level rise. In the early part of Chapter 2, the connection between sea level rise and climate change was discussed in some detail. This connection between the causes of climate change, particularly anthropogenic causes, is important because it helps place prevention policy options into context. As stated in Chapter 2, heat energy exchanges between the systems components of the Earth are the primary means by which sea level rise is occurring; as the Earth retains more heat energy from the sun within its system components, there is a warming trend that leads to the melting of ice and expansion of water within the Earth system. To understand the role of prevention strategies, the proximate cause of the Earth retaining more heat energy must be determined; if possible, stopping the cause of increased heat retention by the Earth would be the ultimate policy goal in a preventative strategy; if the Earth stops getting warmer, then sea level rise should abate.

As shown in Chapter 2, the proximate cause of sea level rise is carbon forcing by human activity that removes carbon from a stored area of the Earth system (the lithosphere) and moves it to the atmosphere where it concentrates and helps form a greenhouse effect. If true, then policy strategies aimed at preventing sea level rise would need to focus on preventing the human activities that are forcing carbon into the atmosphere. This means that policies need to focus on transitioning our energy dependence from hydrocarbons to other sources of fuel for generating electricity, heating our structures, and providing fuel for transportation. A lot of policy work is currently being done on this end. However, the current concentrations of carbon dioxide in the atmosphere, along with residence times of that carbon in the atmosphere, suggest that the Earth will continue to warm for the near to mid term (IPCC 2007). The inability to prevent the continued warming trend over the near term suggests that we must deal with a degree of sea level rise, although preventing further carbon forcing into the atmosphere can be helpful in limiting the amount of warming and thus the worst-case scenarios of sea level rise discussed in Chapter 2. Still, preventative strategies focus mainly on areas outside the direct control of coastal managers and are therefore not a major focus of this text.

Mitigation is about lessening the impacts of an outcome. In terms of sea level rise, mitigation is about ensuring the extent and effects of rising seas are made as negligible as possible. The impacts of sea level rise can be mitigated by limiting the future extent of sea level rise (by limiting the amount of heat energy trapped in the Earth system), and by engaging in actions on land that limit the potential impact of rising seas on coastal communities. The first kind of mitigation, limiting carbon emissions, is a role reserved for other policy arenas. The second kind, limiting the impact of sea level rise through actions taken along coastal regions, fits directly into a number of the adaptation approaches that are discussed later in this chapter. In essence, mitigation strategies that seek to limit the effects and impacts of sea level rise on coastal communities are closely related to adaptation strategies because, at heart, these kinds of mitigation measures look to change the way humans interact with coastal areas based on the effects of rising seas.

An example where a mitigation technique can become part of an adaptive strategy includes the beach nourishment projects mentioned earlier. Beach nourishment can be an important part of a larger set of policy instruments geared toward adapting to rising sea levels. As sea level rises, nourishment projects can aid in limiting the migration of the sea landward. This measure mitigates the impacts of sea level rise and can be used with other policy techniques that look more toward the overall changes needed in adapting to continuous sea level rise over time. The point here is that adaptation techniques can include measures that mitigate the *impact* of sea level rise, but these techniques are not intended to mitigate the *cause* of sea level rise.*

Placing prevention and mitigation of the causes of sea level rise aside and instead focusing on the impacts of sea level rise through adaptation strategies is the primary focus of this text. *Adaptation* encompasses a wide range of approaches to sea level rise that can include elements of mitigation, as described above, and even some elements of prevention, depending on how the term is being used to explain adaptive approaches.† Essentially, the focus on policy frameworks surrounding adaptation is really premised on the idea that sea level rise is happening (although the extent is unknown) and that existing and likely future sea level rise *justifies changing what people do today if the outcome has an expected net benefit* (U.S. Climate Change Science Program 2009). This qualifier, that adaptation approaches make sense *if* the outcome of such approaches creates an expected *net benefit*, needs to be explained in greater detail before moving forward into specific policy approaches of adaptation strategies.

3.2.1 Quantifying a "Net Benefit" in Policy Approaches to Sea Level Rise

Thus far, the policy framework surrounding sea level rise adaptation strategies includes the following propositions:

1. Sea level rise is being observed and there is evidence to support the fact that it is currently occurring.

---

* The cause of sea level rise is carbon forcing leading to heat trapping in the atmosphere. Mitigating the impacts of sea level rise absent consideration of mitigating the causes of sea level rise is the context of what is being discussed in this statement.

† For example, *prevention* of the *impacts* of sea level rise may include armoring techniques such as the building of seawalls to hold back the rising tide. In a sense, this particular approach of holding back the sea "prevents" the impacts of sea level rise by not allowing the waters to move landward. If one sees this measure as preventative in scope (again preventing *impacts* and not *causes* of sea level rise), then the idea of preventative techniques to aid in adaptation strategies is a valid way of thinking about such strategies. The focus here should be on semantics, or understanding the meaning of the term in relation to how it is being used in context.

2. The future extent of sea level rise is unknown; however, upper limits can be estimated and a number of lesser outcomes can be determined based on this upper-limit estimation.
3. The background environmental conditions currently observed on Earth establish limits upon which human society and economic activity are constrained (Figure 3.1).
4. Prevention and mitigation of the *causes* of sea level rise are not part of the framework from which adaptation strategies are derived. However, prevention and mitigation of the *impacts* of sea level rise may be part of the policy framework from which adaptation strategies are derived.
5. Adaptation approaches can include measures that change what people do today if the outcome of those strategies has an expected net benefit.

Previous discussion has explained the first four propositions stated here; what follows now focuses on the fifth proposition of what constitutes an *expected net benefit*. The notion of a net benefit must be reconciled in the framework setting of adaptation approaches so the economic, social, and ecological considerations discussed later in this chapter have relevance in relation to understanding how net benefits might be identified, quantified, and compared, depending on the assumptions made and approach taken in comparing options.

To begin, the term *net benefit* must be placed into the larger context of sea level rise. A net benefit is often judged through a benefit-cost analysis framework, where the proposed benefits of an action are determined, the proposed costs of an action are calculated, and then the benefits are weighed against the costs to see if the action is, on balance, a "net" benefit or cost (Field and Field 2006). Probably the most important aspect of making a net benefit decision is ensuring that *all* benefits and costs are being calculated in the analysis. This includes thinking about the different categories of *values* that can be impacted by a decision, as well as consideration of the impact of a decision at different time frames. For example, a benefit-cost analysis that considers only direct economic variables may come to a different net benefit than a similar analysis considering social and ecological values in addition to direct economic variables. Similarly, a benefit-cost analysis may come to a conclusion about net benefits today that is different from the net benefits that are calculated for a future time. The point here is that a "net benefit" calculation is highly dependent on the factors included in the benefit-cost analysis; failing to include relevant factors in the analysis can lead to a conclusion about net benefits that is incomplete (Field and Field 2006).

One of the variables that impacts net benefit calculations is time. As mentioned above, the time frame considered can alter a net benefit result: Shorter time frames will favor certain factors, while longer time frames may expose other factors that are of equal or greater importance. For example, the decision to allow development along a coastal region based on the current state of sea level may provide immediate benefits that exceed costs; the development allows for immediate economic activity that includes construction of residences along the coastal area, property tax revenues derived from these new coastal residences, and an increase in associated factors that are generally related to greater economic output (tourism, recreation, retail, etc.). However, allowing that same development in an area likely to be impacted by sea level rise over a longer time frame, say 20 years, diminishes the expected net benefits when the impacts over time are included in the calculation.

Time is not the only factor that influences net benefit calculations. Other factors include associated environmental costs from engaging in actions that diminish specific environmental assets over time. For example, coastal wetlands are dynamic areas that provide important functions including direct and indirect values from which humans benefit. A number of these functions are highlighted in Chapter 2, including the following: water filtration, sediment stabilization, habitat for commercially important marine species, and storm protection. There are many other functions provided by coastal wetlands and the value of these functions—direct, indirect, and non-use values—have been calculated to be substantial (Costanza et al. 1997; Wilson and Liu 2008). A policy choice that omits wetland impacts fails to identify important resource impacts that are determinants of "net benefit" in any reasonable definition of the term. In addition, the impacts of a policy direction to such wetland resources needs to include the element of time; what are the impacts to this resource over time? A decision to armor a coastline to protect the high-value residential homes lining the coast may make sense when the cost of building the seawall is measured against the cost of losing those residential homes. However, if armoring erodes an important coastal wetland resource over time, then the costs of losing that coastal wetland should be considered in conjunction with the costs of building the seawall. It may be that, on balance, the "net benefit" over time indicates it is better to forego armoring the coastline and allow the wetlands to migrate with the incoming tide. Other options, including raising, moving, or abandoning the coastal home, may offer a more cost-effective alternative to armoring where the environmental values of the wetland are included in the analysis.

While environmental values are an important consideration when engaging in policy related to sea level rise adaptation, it must be remembered that policy development and implementation are influenced by the political process, which means that social factors play a role when making choices about policy directions. Take the above example about including the value of wetlands into a consideration of whether or not to armor against the incoming tide. The suggestion made above was that a policy decision that included the value of the lost wetland could outrank the protection of the residential properties that lie along the coast. Think about the political fallout from such a decision; essentially what is being stated here is that government will favor a non-human interest over a human interest. Consider this from the standpoint of the residential homeowner who has committed substantial personal resources to the building, maintenance, and associated costs involved with the property in question. Moreover, the private citizen engaged in this conduct because he or she was allowed to do so by existing government policy.* A suggestion that government would allow the development to occur only to "abandon" the private citizen when circumstances change seems to be an ill-advised policy approach, and in fact a policy approach that does not match how many view the role of government as a support system to private actions. The point here is that objective policy analysis should certainly include the addition of all known costs and benefits to a proposed action. However, the weighting of those relative benefits and costs may be altered by factors that are not necessarily objectively based, but rather derive from preferences that rank order values in less objective ways.†

It is certainly true that political and social considerations, as outlined above, impact the objective nature of policy decisions related to sea level rise adaptation strategies. However, there is one factor that tends to influence adaptation strategies more than any other: *uncertainty* over future sea level rise. Uncertainty matters because it prevents absolute knowledge, and a lack of absolute knowledge will always tend to discount the impact of a future event.‡ Not knowing what will occur in the future provides a reduced level of certainty as to the impacts. For sea level rise, the uncertainty of the extent of sea level rise tends to discount actions taken today that would protect against future sea level rise, particularly where actions

* Existing government policy includes incentives to spur private development of coastal areas in order to maximize the direct economic gains associated with coastal resources. Historical examples of government policies used to encourage coastal development include the public funding of infrastructure development (roads, bridges, utilities, sewage systems, etc.) to provide the foundation for private development. In addition, there has been a history of publicly subsidizing the risk of coastal hazards to encourage private investment through cost-sharing of insuring private properties against coastal risks; premiums paid by coastal homeowners often do not reflect the full risk associated with coastal development as some of that risk is absorbed through public assumption of a portion of that risk.

† The fact that environmental values play a role in long-term human well-being may be discounted in political settings, thus allowing the values associated with the coastal wetland to be outweighed by the immediate interests of coastal homeowners demanding assurances that their properties will be protected against future sea level rise. Uncertainty over the extent of sea level rise may bolster a desire to satisfy the immediate needs of the populace even where evidence may show that providing peace-of-mind through creating buttresses today can have net negative consequences for future generations. The equity issues in political decisions where current interests are placed ahead of future interests is an oft-debated area of political theory, particularly when applied to issues surrounding the environment.

‡ As a general rule, we tend to discount future events when we are choosing between alternative actions today. Cigarette smoking may increase the likelihood of lung-related diseases in the future, but that likelihood is discounted substantially in the present; thus, I choose to smoke even when I understand the risks involved. The lack of certainty is an important factor in discounting future events; one reason I may discount the potential impact of smoking today at some point in the future is because I am not absolutely sure of getting a lung-related disease; I may be one of the fortunate ones who escapes unscathed my entire life. Also, the existence of intervening events (I may die from a car crash before I die from smoking-related illness) plays a role in discounting future events, regardless of the likelihood of those events occurring.

taken today are expensive and only effective if sea level rise reaches a certain level in the future; an investment that only pays off if a future, uncertain event occurs is a risky investment by any measure.*

3.2.2 The Impact of Uncertainty in Developing Future Sea Level Rise Policy Frameworks

Uncertainty dominates most discussions associated with sea level rise adaptation. How much will sea level rise? What probability exists for different levels of rise? And what impacts would occur locally under the different scenarios? These are all viable questions that help to formulate the policy debate locally about what actions to take regarding sea level rise. In sum, uncertainty impacts the question of whether spending money today to reduce current and future risks makes sense (Tietenberg and Lewis 2010). This question is really not about decisions related to current sea level rise where the level of uncertainty can be bounded into a range of likelihood for decision-making purposes. Rather, this question is about uncertainty as to the *extent* that sea level rise impacts a *proactive* adaptation philosophy, particularly when the issue is being viewed under a strict economic lens when determining net benefits. Thus, to create policy frameworks that include proactive options, the uncertainty of future sea level rise must be resolved. There are a number of ways to do this. For example, a discounting method can be used to quantify the level of uncertainty into a decision-making process. Alternatively, policy options can be developed that create "net benefits" regardless of uncertainty; the options provide discernable net benefits whether or not sea level rise occurs in the future. Uncertainty can also be resolved when all the costs of a potential action are internalized by the actors involved, thus allowing transparency to help foster better decision making by understanding the full tradeoffs between policy responses. These example methods are outlined now in order to provide a sense of how policy can be framed around the uncertainty involved with future sea level rise.

Discounting is a tool that is often used in economic analysis to determine the present value of something given or received at a future date (Tietenberg and Lewis 2010). Spending money *today* to protect against sea level rise that may or may not occur *tomorrow* would be an example of expending present resources in order to prevent a potential future risk. When money is spent today on a benefit for the future,

* Such an investment may be warranted in situations where the impact is shown to be high even where the probability of the occurrence is low. Fail-safes are often implemented for high-impact, low-probability events such as a meltdown at a nuclear facility. The cost of the fail-safe is justified, even though it may never be used, where the consequences of the event are substantial enough. In such cases, the up-front costs are considered insurance policies of a kind, where the cost is associated with diminishing a risk, even where the probability of the risk is low.

that money is no longer available for other things today.* Whether it makes sense to spend money today on a future potential has to do with the degree of uncertainty surrounding the subject; as uncertainty is reduced, the commitment of resources may make more sense because there is a greater likelihood of the event occurring. To place this in the context of sea level rise, a more rational policy approach may be to commit resources toward sea level rise only to the extent that seas are actually rising: As more sea level rise occurs, more resources can be committed. For example, the choice to armor against incoming tides to protect residential housing along the coast may not make sense when the sea level is well below the level of the homes; this may be true even where the homeowners desire the seawall as a "symbol" of long-term protection to reinforce their commitments to their individual seaside properties. If sea level rises over time in a way that threatens the coastal properties, then armoring may make sense at that time as the uncertainty regarding the extent of sea level rise has been reduced by a wait-and-see approach.† If the properties have been abandoned or destroyed for other reasons during the intervening time (say, consecutive hurricanes have wiped out the area), then the resources that would have been committed to the armoring are available for other uses.

Discounting allows policy choices to be framed in an approach that differentiates between values based on the level of uncertainty surrounding the risks posed by sea level rise. Where uncertainty is high, the level of risk is unknown. Where uncertainty is low, the level of risk has a greater chance of being known. This relationship is shown in Figure 3.2.

Viewing this relationship as a continuum, one can see how the likelihood of a risk becomes clearer as uncertainty lessens, establishing an inverse relationship between a known risk and the uncertainty surrounding an issue. In sea level rise, increased certainty about the risks posed from current and future sea level rise can help formulate the precise value associated with a set of policy response options. Thus, in discounting the value of current expenditures for future benefits, policy frameworks can be established where the level of risk associated with future events correlates to the amount of uncertainty corresponding to that risk. As information about the certainty of the risk improves, uncertainty is diminished, allowing the value of committing resources today to a future risk to be made with greater certainty. The overall value from a policy standpoint is to ensure that decisions about sea level rise are being made with the best available information and

* Making the decision to spend money today irreversibly commits that money for that particular purpose; once spent for one thing, the same money cannot be spent on something else. The availability of money for other uses is often referred to as *option value*. In instances where uncertainty is high (one does not know what is going to happen), it is often argued that option value should be maintained, thus suggesting resources should only be expended when uncertainty is reduced.

† This is not meant to suggest that armoring is the superior policy choice in such circumstances, but rather only to show how option value is maintained while uncertainty is being reduced.

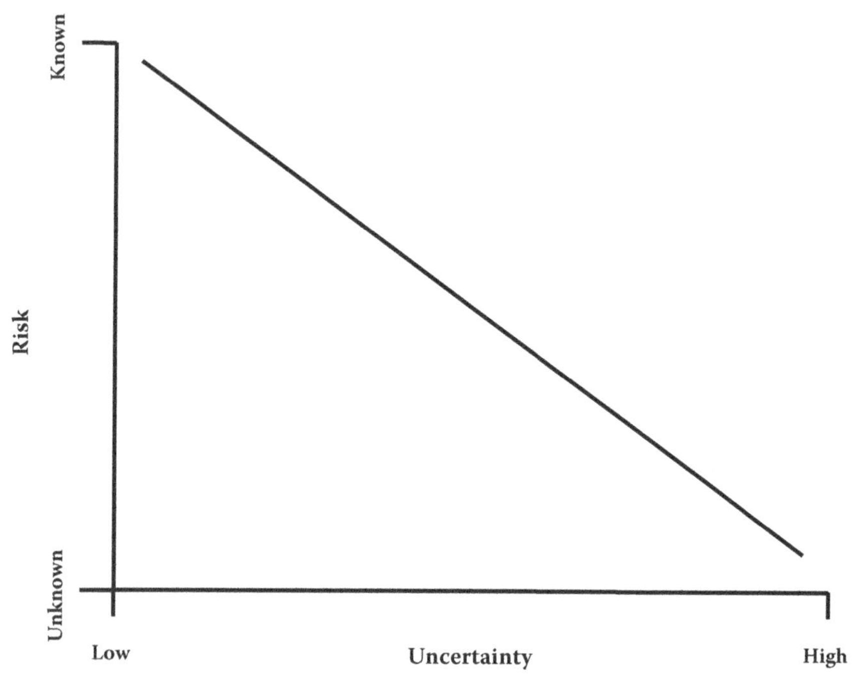

Figure 3.2 Relationship between uncertainty and risk.

requiring a commitment of resources in cases where there is strong evidence that the commitment will yield an overall net benefit.

Using discounting as a way of analyzing policy alternatives through comparing net benefits *between* policy options helps create a process-orientated policy tool that informs decisions between alternatives. A policy option that prioritizes certainty ensures resources are committed today that have a greater likelihood of providing benefits at some time in the future. Another approach using discounting is to look at the goal of the policy to see if a particular policy option can serve multiple objectives, some of which contain guaranteed benefits today while also protecting against a likelihood of risk at some time in the future. This kind of tool is being referred to here as a *win-win* policy orientation, identifying a conceptual framework that focuses on *both* short-term and long-term goals.

Policy options that provide for immediate benefits while also hedging against the possibility of future events are preferred under a win-win framework because they ensure that resources are being utilized in a way that provides immediate value today while also having the potential to provide value over the long term should certain events come to pass. The fact that immediate value is being provided today mitigates the uncertainty of future sea level rise, including the commitment of resources under uncertainty. Because immediate benefits are obtained, the commitment of resources is justified; future sea level rise will provide *additional* benefits toward the initial commitment. Mathematically, if the investment today is less than the immediate benefits achieved from the investment, then the cost is worthwhile. However, if the investment is *both* less than the immediate benefits *and* it may also provide additional long-term benefits with no additional up-front cost, then the investment can be judged a rational expenditure of resources. Win-win policy options are ones that seek to provide immediate benefits that exceed up-front costs, while also identifying potential long-term benefits that accrue in the future should certain conditions occur without any additional investment (cost) required.*

Applying a win-win framework to sea level rise includes the identification of policy goals that provide an immediate benefit in coastal regions while also having the potential to protect against future sea level rise. As indicated above, immediate investments tend to make sense where the cost of the investment is equal to or exceeded by the immediate benefits *and* there is also the potential for benefits to accrue into the future (Nordhaus 2001; U.S. Climate Change Science Program 2009, p. 141–156). The kinds of investments that tend to fall within these guidelines of benefits include the following:

- There is a large risk of near-term harm, and that risk can be reduced by expending resources immediately.
- There is a small cost to reduce the risk.
- The expenditure of resources today shifts the risk from today to some time in the future.

Instances where there is a large risk of near-term harm that can be reduced immediately and where the cost to reduce the risk is small are both cases that require the expenditure of immediate resources; they both incur immediate costs. The third category, reallocating/clarifying risks from sea level rise, does not incur immediate costs and thus the expenditure of resources. Rather, the costs are incurred at a future time as the policies are all geared toward actions today that

* A situation may also be win-win wherein the initial investment (cost) is *equal* to the initial benefit, but the investment also contains the potential to have long-term benefits realized with no additional costs. Further, it may be argued that a win-win may also occur where initial investment is somewhat *less* than initial benefits achieved, but the investment has a strong potential for achieving long-term benefits that outweigh the small initial "net cost" up-front. For example, the reduction of risk alone for future harm may be enough to justify expenditures today to mitigate the potential of an unknown (but possible) risk in the future (insurance is one example of such a strategy).

establish costs that may be incurred in the future; the costs are triggered by the event of sea level rise (and associated phenomena) occurring.*

A large risk of near-term harm includes a variety of coastal zone management activities where the expenditure of funds has an immediate impact that usually meets or exceeds the funds expended. Policy techniques that help to prevent a large risk of near-term harm include beach nourishment projects, wetland protection, home elevation, and dike fortification, among others. Examples where there is a small cost to reduce the risk include: building setback requirements along the coastal zone; new home construction requirements, including minimum elevation requirements and mobility designs; larger coastal drainage system requirements; and infrastructure design elements including higher elevations for road building and bridge design. Finally, examples of reallocating or clarifying risks from sea level rise include: removing subsidies for coastal development; land-use regulations that allow public rights to move with rising tides; shifting the risk of damage to the private landowner through removal of public insurance subsidies/higher insurance rates; and adopting policies that stop shoreline protection activities (U.S. Climate Change Science Program 2009, p. 141–156).

The point of a win-win strategy in thinking about policy goals related to sea level rise is to further bridge an understanding between the relative costs of adaptation strategies to the spectrum of benefits that can be obtained by different policy approaches. Remember that what is driving a lot of this additional analysis is the degree of *uncertainty* that exists about future sea level rise. If there was no uncertainty about the future extent (and thus impacts) of sea level rise, then the decision making could be almost straightforward in many cases. However, the lack of certainty makes decisions about the expenditure of resources toward adaptation strategies more difficult because the perceived "net benefits" achieved through a decision might be difficult to quantify, particularly where the benefits are meant to accrue at some time in the future. The adaptation techniques identified above are examples of strategies that can achieve immediate benefits (addressing large, near-impact events and low-cost preparation) or shift the cost of adaptive strategies to future years where the cost is ultimately incurred by the event occurring (reallocating/clarifying risks). In both cases, net benefits are more clearly articulated because immediate benefits are achieved or the costs of adaptation are shifted to a future time when the risk of sea level rise is better understood.

Internalization of costs is yet another method that can result in better decision making. In general, people make more informed decisions when all the costs of a

* Of course, there are "costs" involved in a change in policy direction. Even if there is no immediate *expenditure* of resources, one can still incur costs associated with changing policy preferences. For example, a policy that removes government insurance subsidies for armoring and insurance along coastal areas adopted today may not incur immediate expenditures of money; however, the policy can limit private investments along the coastal area, resulting in less tax revenue and economic activity, which is a "cost" that is immediately incurred and may or may not be proven appropriate over time (depending on future sea level rise and associated factors).

decision are internalized (Samuelson and Nordhous 1989). The basic idea behind an internalization of costs policy strategy is to determine where costs are borne, identify how the bearing of costs creates incentives among participants, and then attempt to determine the effect of reallocating costs to see if that might aid in achieving a particular policy goal. Consider a situation where a person purchases an insurance policy to guard against a high-risk event. The person may not be willing to engage in the high-risk event if the *full* cost of the risks were internalized to that single person.* However, because the person has obtained an insurance policy to protect him/herself from the high-risk event, that person might be more willing to engage in the event; because the risks are shared, the person internalizes less of the risk leading to a decision-making process that may be different if that person were forced to unilaterally internalize all the risk associated with the event. This is the premise behind policies that force the internalization of costs; by forcing a person to bear the full costs associated with a particular risk, that person might be less inclined to engage in the risky activity. In sum, the internalization of costs can help resolve the issues of uncertainty, particularly where the uncertainty might require the expenditure of public money today in order to hedge against a possible outcome in the future. Rather than expending resources, cost shifting allows a public entity to move some or all of the risk associated with an uncertain future event onto another party.

In coastal management, the internalization of costs can be accomplished through policies such as those described earlier, to include the reallocation or clarifying of risks. Coastal development is often a shared enterprise where both private and public financing is used to develop and maintain a coastal existence. As stated in Chapter 2, the main reason for this is that coastal areas are dynamic and subject to powerful events (e.g., storms) that can lead to devastating consequences (e.g., flooding). Historically, public incentives were an important part of settling coastal regions (Patterson and Glavovic 2008). Today, coastal areas are still dynamic and maintain the dangers normally associated with such regions. However, they are also some of the most expensive real estate areas in the United States, a price signal indicating that strong incentives exist from both the public and private sectors to allow for new development while maintaining existing

* This depends on the degree to which the person *discounts* the risk. Where the risk manifests itself in the future, there is a greater chance for discounting—such as smoking today even where it increases the risk of lung disease. However, to the extent that the person is responsible for internalizing the full costs of the activity, there is a greater likelihood of that person making a decision based on those risks particularly as those risks become realized. A person warned by a doctor of an elevated risk for a heart attack because of diet may be more willing to change habits than in the past as symptoms of the risk begin to show themselves. This is an example where internalization of costs can alter behavior before the full risk is realized.

coastal development.* Current and future sea level rise adds another dimension to this backdrop as dynamic beachfront activity increases. Coastal managers may choose to reallocate risks as a means of limiting the uncertainty related to future sea level rise and its associated impacts.

Reallocating the risk of loss to the private landowner can help create incentives where private coastal landowners are required to make decisions based on a complete internalization of the hazards presented. For example, land-use planning that shifts the risk of future sea level rise solely to private landowners can provide incentives to change private behavior patterns. An owner wishing to develop coastal property may be less willing to include structural and placement considerations for the home that are designed to mitigate against future sea level rise when the risks of loss can be externalized by the private landowner. The kinds of structural and planning capacities to mitigate coastal damage include placing the site of the building as far landward as possible away from the ocean to plan for future sea level rise, including elevation of the building into the construction design to mitigate sea level rise and flood impacts, and designing the building in a way that it can be easily moved away from future sea level rise. Providing for these additional capacities in the development of a coastal residence adds incremental costs to the project (U.S. Climate Change Science Program 2009, p. 141–156). However, a private landowner may be more willing to engage in these actions where they know up-front, for example, that there will be no opportunity to armor against the rising tide; a public zoning ordinance that prevents armoring, either publicly or privately, in the area allocates the risk to the private landowner. Contrast this situation with one where the private landowner assumes the capacity to armor at some point in the future. This up-front belief can provide a disincentive to plan for sea level rise today by the owner because they are able to discount the future event of sea level rise through risk-shifting considerations, including the capacity to deal with the future event through armoring.

Risks can also be shifted from public to private parties through insurance adjustments. Insurance policies can more accurately reflect the true costs of living along the coastline, and that cost can also be redistributed fully to the private coastal homeowner. Public flood insurance programs have long subsidized the cost of premiums paid by private coastal landowners (Hayes, Spafford, and Boone 2007; Hayes and Spafford 2008). Insurance itself is a way of shifting risk, as the

* Private parties have a strong desire to maintain coastal property ownership because of the high monetary values associated with coastal property. Public parties, particularly local governments, have additional incentives to encourage coastal development, also because of the high value of coastal property, as this value reflects higher property tax bases for the community without, in many cases, a proportionate increase in public service responsibilities—thus achieving a "net" gain for the local government. Of course, from a pure "risk" basis, the high value of coastal property may be difficult to justify when the constraints of analysis internalize the additional risk of coastal living, particularly in an era of sea level rise, while dismissing the irrational demand for coastal living.

premiums paid to insure a property from loss generally are far less than the actual replacement cost. However, publicly subsidizing the risk of loss is a further incentive for private coastal landowners to develop along the coast because the actual risk is further diluted in the eyes of the private landowner through the public subsidy provided.* Removing these subsidies provides some basis for coastal landowners to internalize the risk associated with living along the coast. In addition, adjusting the premiums over time to cover the increasing risk due to sea level rise provides reinforcement to the private coastal landowner of the actual risks associated with living along the coast. Shifting this risk can provide for incremental cost adjustments borne by the coastal property owners and the resulting choices made can more accurately reflect the actual risk involved with coastal development. Less coastal development in marginal areas can ensure open space that allows for the realization of other values (such as wetland migration), while insurance premium credits for proactive mitigation measures like structural enhancements to coastal development (a lower premium where the building is elevated, set back from the shore, or built to be easily moved) can reinforce sound building practices that mitigate the impacts of future sea level rise to the coastal community.

A few additional points of consideration should be mentioned with regard to this concept of reducing uncertainty through the shifting of risk to private coastal landowners. First, it must be acknowledged that socio-economic status will impact the capacity to deal with risk shifting because, ultimately, the shifting of risk is really tantamount to increasing costs associated with the given activity; those who can afford the additional costs through the shifting of risk may continue their activities, regardless of the increased costs associated with the activity. For example, wealthy landowners in prime coastal areas may discount the additional premium associated with insurance rates because the costs are marginal in comparison to their overall wealth. In addition, the incremental additional cost in insurance premiums may be offset and overshadowed by the additional value attached to the property existing on the coast; an additional insurance premium of thousands of dollars is relatively small in comparison to a coastal property than can be hundreds of thousands to millions of dollars more valuable than similar properties in structure and land area that exist inland from the coast. Because of the relative value of money to those of different means, the policies where risks are shifted onto private

* Publicly subsidizing the risk provides an additional benefit to private landowners along coastal regions. As previously stated, coastal land is some of the highest value land because there is generally strong demand for coastal property at prices that exceed similar properties located inland from the sea. Any form of subsidy can reinforce a private choice to develop along a coastal area because there is already an inflated value associated with coastal areas to begin with. Thus, there is an existing incentive to develop in a coastal area because of the premium often paid to live along the coast; the additional amount willing to be paid for the coastal property already insulates the private coastal developer/owner from some of the additional risk that exists along the coast. Any public subsidy provides an *additional* subsidy to the already-existing subsidy of the higher value attached to coastal properties.

parties must also be viewed in terms of their impact on the affected individuals as equity considerations abound. While government may opt to shift certain risks to private parties as a means of encouraging proactive approaches to sea level rise, the impacts of such policies on different classes of citizens must be considered.*

To summarize, this section highlighted the importance of *framing* policy questions related to sea level rise. Important considerations here include the quantification of a "net benefit" in policy approaches to sea level rise, and also the impact of uncertainty in developing policies that attempt to address future sea level rise by acting today. The "net benefit" calculation is one where expenditure of money today provides sufficient benefits that equal or exceed the costs incurred. The clearest way to accommodate a "net benefit" approach is to identify policy approaches providing immediate benefits that equal or exceed immediate costs *and* also have the potential to provide additional benefits in the future without any additional costs. Uncertainty plays an important role because the greater the uncertainty about a future event, the less probability there is that current expenditures will lead to future benefits; the uncertainty makes the likelihood of future benefits unclear. Providing immediate benefits is one way to mitigate this problem when considering policy approaches. Examples of coastal management techniques that have immediate costs but provide immediate benefits were presented in this section. They include beach nourishment projects, wetland protection, and associated strategies. Another mechanism is to enact policies that do not have immediate actual costs, but rather incur the costs over time as the level of uncertainty about sea level rise diminishes. Policies that shift or reallocate risks to private coastal landowners today can help in mitigating the costs of adapting to sea level rise in the future. Examples include land-use policies that remove the assurance of protection against future coastal hazards, as well as policies that remove subsidies for insuring coastal properties. Over time, as sea level rise hazards become realized, the "costs" associated with coastal living increase, and those costs are internalized on the private coastal landowner through risk-shifting policy approaches. Although discounting may occur in how those costs affect the coastal landowner, particularly for wealthy individuals,

* This point can be made by analyzing the marginal value of a dollar, and comparing its marginal value to people with different financial capacities. A dollar is always worth a dollar in real terms, meaning the value of the dollar is established and only fluctuates based on a comparison of its value in relation to what it can be used to purchase (often referred to as *inflation*). However, the value of a dollar differs depending on the relative wealth of the person in many instances. For example, a person who is rich has many dollars; thus, the marginal value of a single dollar to a rich person would be less than the value of that same dollar to someone who is poor (because the poor person has fewer dollars). In this sense, a policy that shifts costs onto private individuals can have different impacts, depending on the relative wealth of the person under consideration. A wealthy person might not be as affected as a poorer person by having to pay additional dollars through a risk-shifting policy measure. It may be argued then that the impact of a risk-shifting policy through internalizing costs can be inversely proportional to the wealth of the individual; the impact of the policy is greater to a poorer person, whereas the impact of the policy is lessened to a wealthier person.

the goal would be to limit the amount of risk aggregated for future sea level rise or, at a minimum, create an environment where the expectations for accepting those risks are placed solely on the private landowner.

Now that a policy framework has been established to understand some of the constraints involved in planning for sea level rise adaptation, a review of the general categories of adaptation approach choices follows. The adaptation approaches summarized below should be viewed in light of the framework mentioned above and further evaluated based on the economic, social, and ecological considerations that follow in the final section of this chapter. Remember that any single option described below must be viewed from a variety of perspectives in order to understand the potential impact that option has on those perspectives. A choice that seems reasonable from an economic standpoint may have strong social implications; requiring higher insurance premiums may make economic sense but may disproportionately impact people of lower socio-economic means. Similarly, a choice to armor may seem rational when protecting valuable coastal assets that preexisted sea level rise, but doing so may harm important ecological values that, on balance, indicate that armoring is not a viable policy option. As new material is presented below, make sure to bring along the concepts previously discussed. The frameworks identified above should be viewed in light of the scientific principles summarized in Chapter 2, and this collective set of information should form a conceptual framework as you consider the following materials.

3.3 Adaptation Approaches

As sea level rises, coastal communities are presented with a set of choices that are constrained based on the circumstances presented in the particular coastal area under consideration. Chapter 2 presented a variety of factors that can influence the overall impact of sea level rise, such as existing development and geomorphological characteristics. In addition, the first part of this chapter has provided general policy frameworks to help understand some of the constraints involved in making choices about how to deal with sea level rise under the local conditions presented. Combined, the physical factors and policy frameworks help to develop the kind of decision-making approach that might best suit the adaptation strategies chosen. The purpose of this section is to establish the general categories of approaches available in adaptation strategies at the coastal level.

Before moving into the categories of strategies described below, it is important to reiterate that the focus here is on *adaptation* to the phenomena of sea level rise and not the prevention or mitigation of the *causes* of sea level rise as elaborated earlier in this chapter. The adaptation approaches outlined below all assume that sea level rise is occurring and will continue to occur for the indefinite future. In addition, the strategies that follow assume that the coastal areas under consideration will be impacted in some measure by sea level rise. The conditions of a particular coastline that would mitigate or obviate the need for the kinds of responses identified below were discussed in Chapter 2. Some of these conditions may be repeated below to provide context in understanding a particular adaptation approach, and will be incorporated into the subsequent section that follows when considering economic, social, and ecological factors among the different adaptation approaches. However, the goal of this section is to lay out adaptation approaches in a more descriptive sense, providing an accounting of the kinds of approaches available to aid in coastal management decisions without necessarily prioritizing or distinguishing between approaches based on conditions found at a particular coastal area.

As an overview, this section begins with a discussion of coastal adaptation strategies, distinguishing between *developed* and *undeveloped* coastal areas. The purpose is to identify adaptive strategies that are triggered based on the degree of development in a region and how undeveloped areas provide special policy issues for consideration. The remainder of this section is divided into two main categories, both of which assume development of the coastal area under consideration. The first category identifies adaptation strategies where the focus is on *staying* in the coastal area that has been developed. Two main choices are presented: (1) *protecting* against the incoming sea, or (2) *adapting* to the incoming sea. Protection is distinguished from adaptation in the definitional sense that adaptation provides for passive measures where human actions respond to the incoming sea, while protection is meant to reflect an active set of policies aimed at preventing the encroachment of the sea landward. The other main category deals with *retreat*, which is distinguished from staying in the coastal area by emphasizing the strategic removal of human habitation and settlement from the coastal area as the sea moves landward. Although the strategies are divided categorically between staying and retreating, they should not be seen as mutually exclusive. Rather, a combination of policy tactics that involve elements of approaches identified under the "stay" category of options may be mixed with options that include "retreat." As you review the categories below, consider how they might be utilized collectively to achieve the larger policy goals of adapting to sea level rise based on the considerations identified above in the framework section of this chapter (Section 3.2).

3.3.1 Considerations Concerning Developed versus Undeveloped Coastal Areas

Before discussing specific adaptation approaches along coastal regions, this section reviews the special considerations that occur between developed and undeveloped coastal areas. Coastal regions that have a legacy of development present adaptation challenges that are not necessarily presented in undeveloped coastal areas. For example, commitments of human capital, both public and private, have been made in developed coastal areas; as a result, these commitments often limit the scope of adaptation strategies available. In addition, developed areas immediately adjacent to the coast may already have engaged in certain strategies to mitigate impacts to the coast (hard armoring, for example) that also may limit the kinds of options available in nondeveloped areas. A few of the considerations that apply between developed and undeveloped areas are outlined below. The information provided should supplement the later sections on specific adaptation strategies.

Patterns of coastal development can include erratic patchworks where concentrated areas of development are interspersed among areas of relatively undeveloped coastal land. There are a variety of factors that influence this pattern. For example, regions of the coastline that are directly exposed to open ocean may be preferred areas for development. Inner bays and other coastal landmasses that are not directly open to the sea may be areas of secondary preference when it comes to coastal development. As such, areas that have not achieved a consistently high demand of intensive coastal development may show a patchwork of development favored around coastal areas exposed to the ocean with inner bays and estuaries having less development, depending on the coastal features presented (Kleppel, DeVoe, and Rawson 2006).[*]

Coastal areas may exhibit a high degree of development in areas that are proximate to urban centers, where the intensity of development diminishes with increasing distance from the urban center. This relationship is often shown in density mapping, where a preference for living within close proximity to urban centers can be observed, particularly on the northeastern and mid-Atlantic shorelines of the United States. Historically, urban centers provided the focus of living until a variety of factors influenced the decision to move away from urban centers.[†] Coastal regions near urban centers provided idealistic communities where a suburban lifestyle could be maintained within close commuting distance of an urban center where jobs were often plentiful (Kleppel, DeVoe, and Rawson 2006). The kinds of qualities associated with coastal communities mirrored those mentioned above, where direct ocean views and proximity were valued higher than estuary or backbay areas. The resulting trend is to find greater development of coastal areas that are close to urban centers, and less development along coastal areas moving away from the urban center.

The kind of coastal development observed in different regions also varies depending on the time in which the coastal area was developed. Older concentrated coastal developments tend to reflect preexisting choices regarding protection

---

[*] For example, coastal regions that include rivers emptying into a coastal bay area contain a multitude of land–water interface features. Traditional coastal zone areas where sandy beach abuts an intertidal zone of an open bay or ocean are generally preferred over marshland or riverine estuaries that help drain water from the land to the ocean.

[†] Factors include industrialization, highway construction, and the accompanying urban sprawl that influenced suburban developments after World War II. In addition, influences such as "white flight" phenomena observed in major cities during the school desegregation era beginning in the late 1950s also provided the desire to move away from urban centers.

from coastal hazards. Hard-structure armoring is often observed in older coastal communities; seawalls, jetties, and groins are fixtures that can be regularly seen along the Atlantic seaboard from the Carolinas up through New England (U.S. Climate Change Science Program 2009, p. 43–56). These preexisting structures may limit policy options, for example, removing the possibility of allowing the landward migration of the sea (depending on the extent of the armoring), or alternatively reinforcing armoring as the policy of choice for these areas. The unique circumstances encountered can aid in identifying the kinds of policy options available. Three examples follow that highlight how different conditions observed in a particular coastal region impact the policy options that are available to address both existing and future sea level rise.

Both the quantity and intensity of developed coastal land constrains a number of policy planning tools. A fully developed coastal area such as Manhattan Island cannot simply be placed aside and sea level rise allowed to migrate landward. First, there is a substantial commitment of resources to this landmass; large skyscrapers dominate the land and substantial human capital has been placed into its development and maintenance. If simply left to the devices of sea level rise, there would be an irreversible loss of value associated with the area (infrastructure, buildings, commerce, etc.). There is also a substantial symbolic impact associated with allowing sea level rise to overtake a coastal landmass like Manhattan. Finally, because of the intensity and extent of development of the island, there are diminished ecological values that may justify policy options that rank ecological values higher than human-based values. In essence, the circumstances encountered in Manhattan strongly support a policy preference that would fall into a *stay and armor* approach.

A rural coastal area of Maine (as a different example) may provide more policy options pertaining to sea level rise response. Consider a coastal bay town in a rural part of the state of Maine that exhibits a variety of physical characteristics, including a town center that sits near the mouth of a river leading to a coastal bay where human development and habitation are highest. From this town center the coastline exhibits a number of features that are indicative of watershed areas, with small sections of high bluff balanced against low-lying estuaries allowing rivers and streams to converge and dump their contents into the local bay waters. A schematic representing these characteristics is provided in Figure 3.3.

The policy options available here are varied because the area presents small pockets of intense development surrounded by coastal areas where development is limited to nonexistent. Equally important is the fact that the areas surrounding the town center are low lying, thus allowing a migration path for sea level rise that can mitigate the impact on the clustered town center. Policy options here might focus on providing some level of protection for the town center while allowing adjacent low-lying coastal areas to remain undeveloped so as to provide a migration path for the rising seas. This *overflow capacity* provided through open coastal space would allow the town to assess the extent of sea level rise as it occurs, freeing

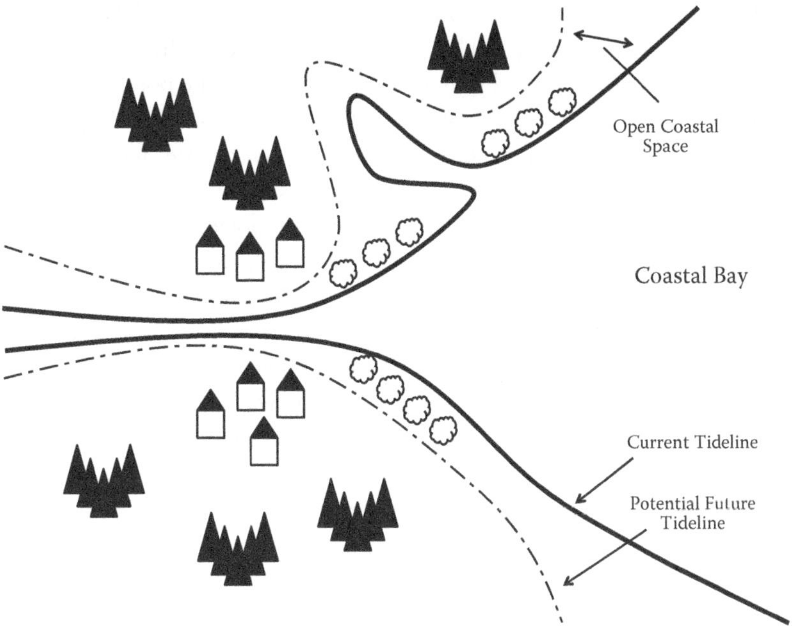

Figure 3.3 Representation of coastal bay town in rural Maine (United States).

up additional policy options for the town at a future date. For example, continued sea level rise may highlight a planned retreat option for the town center, depending on the extent of sea level rise observed. However, by reducing development of surrounding coastal areas, the town provides itself with policy options that allow for sea level rise in the future while also mitigating the impacts of that rise on currently developed regions.

The city of Boston and its surrounding communities provide the third example highlighting policy options for sea level rise related to the question of developed and undeveloped landmasses. Boston is a coastal city that has been developed over time through a mix of settlement and filling of adjacent bay area (Kennedy 1992). Boston has three major watersheds serviced by three primary rivers: the Neoponset, Charles, and Mystic Rivers. Both the Charles and the Mystic Rivers have had their natural watershed areas filled and developed, while the Neoponset still maintains remnants of its lower watershed region due to a lack of damming opportunities due to its relatively slight slope near the river's outlet to Boston Harbor. The development of the city proper as well as the substantial development around the Charles and Mystic Rivers provides little opportunity for natural migration of sea level rise landward over time. Rather, these areas have been buttressed against the incoming sea through a series of filling operations and associated projects that leave armoring or retreat as the major policy options. However, the existence of the lower watershed area of the Neoponset provides a potential source of sea level rise accommodation for parts of Boston and its surrounding suburbs. The existence of the undeveloped and open watershed area can help mitigate sea level rise in the southern areas of Boston and surrounding land. From a policy standpoint, the potential adaptation strategies to sea level rise might include a combination of armoring techniques along with preservation of open coastal space. In areas that are significantly developed, armoring may be the superior option. However, some mitigation of sea level rise can be had by allowing waters to flow into the lower Neoponset watershed basin, lessening the flood potential in other areas of the city. The fact that the Neoponset River may act as a buffer to sea level rise in an otherwise developed region highlights the importance of even small tracts of open space along coastal corridors. Small interspersed areas of open coastal space can substantially mitigate the impacts of sea level rise depending on the relative positioning of those open spaces and their physical characteristics.* In situations where the undeveloped land abuts high-value, developed land, the value of the open space only increases.

The three examples provided above give some indication of the relative importance between developed and undeveloped land as it applies to sea level rise adaptation policy development, implementation, and evaluation. In each example, a different set of policy priorities is presented based on the intensity of human development as well the physical conditions of the surrounding coastal area. Manhattan represents a substantial investment in human development along a coastal area where the history of development has left few policy options. The Maine coastal town example provides many of the opposite considerations, where a mix of low to moderate development with some direct value is tempered by the rural and undeveloped state of the surrounding coastal region; there are more policy options available in this setting than there are for the Manhattan example. Finally, Boston represents a coastal city with both substantial investment and legacy issues that limit policy options, and also some undeveloped area that can mitigate the extent of sea level rise under certain conditions. The existence of the undeveloped watershed area in Boston provides an important policy option that can allow for a number of objectives to be met. In essence, the existence of the undeveloped watershed adjacent to Boston may provide options that are simply not available in a more completely developed area such as Manhattan. Knowing policy options through an assessment of physical assets and local accounting of developed and undeveloped land can help policy planners place sea level rise adaptation options into a larger context that can meet multiple objectives—both today and tomorrow.

* Undeveloped coastal land that is defined by low-lying areas extending landward creates a basin where migrating seas might congregate. The degree to which these open basins exist mitigates the potential impact of sea level rise on adjacent developed areas, particularly when these areas lie above the undeveloped basin or are otherwise armored so they are protected from sea water incursion in relation to the adjacent undeveloped land; the greater the developed coastal area, the greater the value of remaining open space that is proximate to the developed area.

The analysis presented immediately above focuses squarely on adaptation strategies where the priority is placed on values associated with protecting human development. However, it is important to reiterate that human well-being is but one dimension of value when considering adaptation policy options. Other values, including the ecological integrity of the coastal zone, are important considerations when discerning options between developed and undeveloped land. For instance, there may be cases where interspersed human development exists in a coastal region that has critical ecological assets. In these situations, it may make sense to prioritize the ecological values over human well-being and opt for policy directions that favor the landward migration of the coastal zone as opposed to choices that limit such options. The impact on the local citizenry should be considered and can often be accommodated through a variety of options that will be explained in greater detail below. The point here is that human development does not need to be considered the default priority for adaptation strategies to sea level rise; other values and considerations may be equally important. Thus, understanding the full context of the area in question is paramount to beginning any serious consideration of adaptation approaches. With these conditions and caveats in mind, the different adaptation approaches follow.

3.3.2 Stay

The choice to *stay* at the coastal zone is based on a number of factors. The level of development in the coastal area is obviously a primary consideration; places such as coastal cities have invested substantial capital in the area and there will generally be a desire to remain, especially in the early stages of sea level rise (Kleppel, DeVoe, and Rawson 2006). Complicating the choice to stay are the incremental impacts that evolve over long periods of time relative to sea level rise. Aside from worst-case scenarios,* most relevant information on sea level rise indicates that the process is currently slow and relatively steady (Intergovernmental Panel on Climate Change 2007). Because of the slow, incremental pace of sea level rise, the choice to stay in a coastal area can be reinforced; the costs of settling and developing the area have already been borne, whereas the costs of adapting to sea level rise come incrementally over time. Because the benefits of staying (where the "costs" involved in developing the area have been internalized to achieve the benefits of development) are weighed against the costs of dealing with sea level rise incrementally over an extended period of time, the benefit-cost analysis may weigh in favor of staying. Analysis of this kind may create incentives to commit additional resources to staying in the coastal area and adapting to the slow rise of the sea. However, assuming sea level rises to just moderate projections in many areas, there may come

* Worst-case scenarios include high-impact, low-probability events such as a major ice shelf breaking off, moving into open ocean, and immediately beginning to melt at rates that create a substantial addition of water causing sea levels to rise at a faster rate.

a time when the seas have risen to a point where incrementally dealing with the problem is no longer an option. Just how to balance this potential with present decision making is an open question, particularly where the full extent of sea level rise is unknown. However, the current accounting of incremental costs comparing investments already made in developed coastal regions does suggest that staying is a policy option that will often be used in the near term.

While many coastal communities may choose to stay along the coast, the way in which the coastal community chooses to adapt to current and future sea level rise can vary substantially. Categorically, coastal communities can choose to either *protect against the incoming sea* or *adapt to the incoming sea*. Protection suggests proactive policy measures that place barriers between the ocean and the land. The kinds of barriers chosen can vary from those that are longstanding to those that are more impermanent in nature. Adapting to the incoming sea takes a more reactive policy approach where the community adjusts its expectations and actions in relation to incoming sea level rise. The kinds of policies formulated under an adaptation approach tend to focus on allowing the seas to migrate landward while staying in the area, but adjusting the way in which infrastructure and dwellings are built to provide a balance between the rising seas and the choice to continue human settlement of the area. The specific kinds of techniques available under each category, protection and adaptation, are outlined in further detail below.

3.3.2.1 Protect against the Incoming Sea

Protecting coastal areas from the incoming tide refers to a set of human-based activities that is intended to reduce risks associated with sea level rise. These risks can include direct dangers associated with sea level rise, such as inundation of coastal areas. The risks may also include indirect dangers that are associated with sea level rise, including increased storm intensity and duration that lead to flooding and erosion of coastal regions (U.S. Climate Change Science Program 2009, p. 87–104). The kinds of techniques that can be employed to protect against the incoming sea are generally based in engineering, where *shoreline armoring* is used as a means of replacing the natural shoreline with an artificial surface that is then used to hold back the sea (Titus 1998). The types of shoreline armoring techniques that may be used can be divided into two main categories based on the goal intended. One type of technique is *fixed-position armoring methods*; the other technique may be referred to as *flood protection techniques*. The goal of fixed-position armoring methods is to establish a fixed location for the shoreline. The goal of flood protection is to prevent flooding of land when water levels exceed the height of the land being protected (Titus 1998). Both of these categories of techniques have been used in the past to either fix the location of the shoreline for land-use planning, or otherwise protect sensitive inland areas from flooding and inundation that derive from the background dynamic conditions of coastal areas. Sea level rise presents a potential to accelerate the effects of coastal dynamic conditions, and these techniques will undoubtedly be relied upon as methods to ensure the protection of coastal regions in the future. Assessment of factors that include economic, social, and ecological impacts (discussed later in this chapter) must also be analyzed to ensure that the impacts of armoring are part of the decision-making process.

3.3.2.1.1 Fixed-Position Armoring Methods

The purpose of fixed-position armoring methods is to create a set barrier between the land and sea. The benefits of creating an established barrier include the ability to provide consistency in terms of planning for development. Also, the fixed barrier can be built in a way that ensures a level of protection from storms and other sea-related processes (Dean and Dalrymple 2002). There are a number of structures that can be engineered to create fixed barriers, including *seawalls, bulkheads, revetments*, and *retaining walls*. Each method is used to hold back the sea, but the purpose and effect between the methods listed can vary. A summary of each method of armoring follows.

Seawalls are barriers engineered to withstand high-energy impacts from the ocean. The walls are often constructed of hard materials meant to be longstanding, and they are usually designed in such a manner as to dissipate the energy from an oncoming wave. The goal of a seawall is to provide a sturdy barrier that creates lasting protection from the ocean. The result of creating a seawall is usually a stark demarcation between land and sea. Because the barrier is permanent and does not allow for the movement of ocean beyond it, barriers in proximity to tidal ranges often erode the coastal zone and prevent any migration of coastal features landward. Seawalls may be constructed away from the intertidal zone existing upland near a line of vegetation past the dry sand beach. These walls are intended to prevent storm surges that tend to cover the extent of a beach area so the surge cannot move onto portions of land where human domiciles exist. However, the fact that seawalls are generally permanent structures means that walls placed near the line of vegetation today will still prevent the movement of coastal features landward as sea levels rise. Thus, whether the seawall exists at the water's edge or landward of the beach area, they are essentially permanent features that, once constructed, place limits on policy choices in adapting to sea level rise.*

Bulkheads are meant to provide support to nearshore areas. They are walls designed to hold the land in and prevent erosive impacts that often accompany tidal water flows. Unlike seawalls, the purpose of bulkheads is not to prevent impacts from the sea like holding back a storm surge. Rather, bulkheads operate in much

---

* Note that a seawall can certainly be removed but the likelihood of removal once constructed is limited because of the symbolic impact of the wall. A coastal community that develops around the existence of a seawall will likely be less inclined to remove the seawall (protection) as sea level rise occurs. Thus, although the wall can be removed, there is less chance of removal once constructed than when choosing not to construct a seawall up-front.

the same way as a retaining wall; they are meant to offer a degree of support so the land can maintain its relative form and position in relation to what has normally been the land–sea interface. Bulkheads tend to be found away from the open ocean in estuaries and back bays where the tidal energies from the ocean are diminished. They are often used in recreational areas that abut the sea, such as marinas, and they can also be found in residential areas at the water's edge (Dean and Dalrymple 2002; U.S. Climate Change Science Program 2009, p. 87–104). The main advantage of a bulkhead is to provide some stabilization to low-energy marine areas where subsidence of coastal land may occur. From a policy standpoint, bulkheads may represent a less dramatic means of providing short- to medium-term fortification of estuarine shorelines. As such, bulkheads may represent one tool that can offer intermediary mitigation from the impacts of sea level rise, particularly combined with other policy options.

Revetments are like seawalls in that they are built to withstand the powerful energy that is created through wave and storm activity. However, unlike seawalls, revetments are built to follow the natural slope of the existing shoreline. In addition, revetments are often built using structures such as large boulders with multiple faces that collectively act to dissipate the energy of wave action from the ocean. The dissipation of wave energy allows the revetment to provide some stability to coastal zone features, such as a beach or intertidal zone that may lie immediately seaward of the revetment (Dean and Dalrymple 2002). Depending on the location and size of the revetment, the structure may moderate the impacts of erosion seaward of its placement in the near term, while also providing a stabilizing influence for the area that is landward of the structure. From a policy standpoint, revetments can be seen as a potential armoring tool that, like a bulkhead, may be used to moderate immediate impacts of sea level rise without substantially compromising ecological values. This is accomplished by maintaining some integrity of the ecological features that exist in the coastal area seaward of the structure. However, as sea level rise continues, the placement and size of the revetment becomes a concern because a larger structure may limit the movement of coastal features landward, while a more carefully developed structure with interspersed rocky sections may allow for the advancement of the coastal zone landward.

Retaining walls are structures that are built into the coastal zone feature, often being buried within the feature itself. For example, geo-tubes are a kind of cylinder filled with sand that is buried at the landward extent of a beach area (Dean and Dalrymple 2002). The tubes are buried to shore up the landward portion of the beach and create a vertical "wall" intended to provide a kind of sand dune structure to limit the impacts of waves from storm and extreme tidal events. Because the structures are often buried and made part of the coastal feature, they can allow existing coastal features to be retained and move landward. The extent to which coastal features can move landward during sea level rise depends in large part on how the walls are created and the energy dynamics of the area. Unlike seawalls and revetments, retaining walls are not meant to provide a lasting defense against high-energy impacts, but are rather an intermediary armoring option that lessens the effect of high-impact events along a coastal region. Because of their impermanence, retaining walls may be seen as an armoring option that, engineered correctly, may offer near-term mitigation of sea level rise while still allowing coastal features to be maintained over time as the seas migrate landward.

Seawalls provide the strongest degree of shoreline protection at the expense of shoreline attributes. Meanwhile, retaining walls and bulkheads provide a lesser degree of shoreline protection, but often allow many of the existing shoreline attributes to be retained over time. Revetments may provide an intermediary approach with a degree of shoreline protection but also the potential to protect shoreline attributes both today and in the future. Understanding the relative attributes of each of these fixed-positioning armoring methods is helpful when comparing and contrasting between methods to determine policy options that achieve a diverse set of goals in adapting to sea level rise. These fixed-position armoring methods are not the only ways to protect against the incoming sea. There are also flood protection techniques that offer intermediate levels of protection with sometimes a greater range of control over the amount of water allowed to move inland. Flood protection techniques may offer additional tools to a decision maker that mitigate the impacts of sea level rise while also allowing for important coastal functions to continue in some fashion.

3.3.2.1.2 Flood Protection Techniques

Flood protection techniques are primarily aimed at protecting against the intermediate flooding or the permanent inundation of a land area (Dean and Dalrymple 2002). In a coastal setting, ocean storms can cause intermediate flooding of low-lying coastal areas. Meanwhile, development of areas near or below sea level have required engineering techniques that provide for the protection of these low-lying developed areas from coastal flooding. Finally, sea level rise presents the future threat to developed coastal areas, as even slightly higher elevated coastal regions may become at or below sea level in the future, depending on the extent of sea level rise that occurs and the dynamics that factor into the impact of this rise on these coastal regions. The types of flood protection techniques described below include the following: *dikes, dunes, tide gates,* and *storm surge barriers*. Some of these protection techniques differ from one another in terms of the way in which they operate and function, while other distinctions are based more on the purpose of the technique used, which is defined by the surrounding environmental conditions in which the particular flood protection technique operates.

Dikes act much like seawalls in that their intended purpose is to keep the area landward of the dike dry from water build-up (Basco 2003). Unlike seawalls, dikes are usually placed inland some distance from the ocean and are often made from earthen material. Dikes traditionally are used to protect low-lying inland areas from storm surges that may exceed a natural built-up barrier near the shoreline, thus subjecting the inland low-lying area to flooding. Because of the dike's existence, water that accumulates inland from the dike through rainfall and other means must be pumped from the area. Dikes have traditionally operated as a means of protecting low-lying areas from water infiltration and thus they may be used as a tool for preventing sea level rise in the near term. The advantage of a dike is that it is often made of natural materials (soils, etc.) that can be easily removed in a short period of time. So, for example, a dike might be built inland to protect low-lying areas that exist within proximity of the coastal area to bolster near-term defenses against rising sea levels. This may be accomplished to buy time to develop a more holistic strategy for dealing with observed sea level rise. If that strategy includes a planned retreat of the area due to circumstances presented, the dike can be removed after the planned retreat as part of a holistic policy geared toward allowing the natural flow of water to continue to migrate inland with associated coastal features.

Dunes are large mounds of earth material (usually sand) that are formed through both natural and human-made processes; they are normally located on the landward side of a beach area within a coastal zone often existing near a line of vegetation. Natural dunes are primarily formed through wind activity, while artificial dunes can be created through the movement of earth material molded into hill-like structures. Dunes act much like dikes in that they help provide a barrier between the upland areas beyond the immediate coastal zone and the ocean. Dunes can also be important stabilizers of near-ocean sediment, helping to hold sand on the ocean side of a coastal area. In addition, a number of grasses have evolved to take advantage of the niches provided by dunes, growing on either the sea or landward side of the dune system and aiding to stabilize sediment. From a policy standpoint, dunes represent an important means of moderating the impacts of tidal and storm surges. Importantly, dunes also provide an impermanent defense against flooding and inundation, allowing coastal features to move landward under conditions of sea level rise.

Tide gates are human-created barriers that exist in low-lying areas and are used to moderate the influences of tidal ranges. Structurally, a tidal gate consists of two components: (1) the wall structure that elevates the immediate area to hold back the sea, and (2) the "gate" itself, which opens and closes to either prevent water from moving past the gate when closed, or otherwise allow water to flow through the gate when open. As mentioned in Chapter 2, tidal ranges of oceans can have a substantial impact on certain coastal features, sometimes inundating a large area of land during high tide maxima, and then exposing this land during low tide events. Tide gates moderate this process by preventing the influx of ocean water into a low-lying area during a high tide event, and also by allowing water trapped in a low-lying area to escape during a low tide event. The "gate" is the operative mechanism allowing for the control of water inflows and outflows during tide events. When the tide is low but rising, the gates are closed to prevent the infiltration of water past the gate and barrier (wall) that extends from the gate. If water is trapped landward of the tide gate, then the gate might be open as low tide is approaching to allow this water to escape from the area toward the ocean. From a policy standpoint, tidal gates offer a means of controlling water flow through a given access point to areas both landward and seaward of the gate. They offer more options for controlling water than simply preventing water flow.

Storm surge barriers are large-scale versions of tidal gates with a slightly different purpose. While tidal gates operate to utilize tidal ranges, opening and closing based on the desire to move water to and from landward areas, storm surge barriers are used to prevent or mitigate storm surges. The gate of a storm surge barrier is usually open to allow tidal fluxes to move in and out of the barrier system. When a storm surge is detected, the gates are closed to prevent further infiltration of water inside the system. Barrier systems vary in operation, sometimes lowering a gate to prevent water infiltration, and other times raising a gate from the depth of the water to prevent infiltration. These kinds of systems have been created to deal with the inundation that occurs in the coastal area due to storms; they are usually present because the damage from a past storm has justified their cost. Thus, the areas where they exist have generally been proven to contain the characteristics of landmass elevation and population density that justifies their construction. Continued sea level rise will only exacerbate the need for the use of these barriers as a mechanism to hold back the coming tides and possibly even operate indefinitely in a closed state if sea level rises high enough (effectively becoming a seawall).

The different armoring techniques described here (fixed position and flood protection) are both ways of proactively acting to limit the extent of ocean migration landward. The techniques vary in terms of their intended purposes and lasting effect; some are meant to create a precise delimitation between land and sea (hard armoring), whereas others are intended to control the flow of water during specific times and events (tidal gates). Each technique described has attributes that can be ranked based on the goals of a given policy objective. Of key importance is ensuring that the goals of a policy direction are clearly articulated and understood so superior (or inferior) techniques for holding back the sea may be determined in relation to those clearly articulated goals. Failure to properly identify policy goals in this context may lead to the use of a technique for armoring that is inadequate in some measure; for example, the chosen technique fails to protect ecological values when those values are identified policy goals.

The several techniques described here are all related to *protecting* against the incoming sea. This is true even where certain techniques moderate the level of protection offered; some techniques described offer a strong level of protection, while others mitigate tidal impacts but are less capable of holding back the incoming sea—particularly in relation to high-energy events. A slightly different approach for adaptation strategies focused on remaining at the coast is *adapting* to the incoming sea level rise. Some techniques described in this section may be seen as mechanisms that aid in adaptation; for example, by buying the coastal community time in addressing sea level rise through mitigating the impacts of sea level rise; dunes and other structures mentioned are capable of mitigating the impacts of sea level rise in a way that allows for consideration of adaptation strategies. Below, a number of adaptation-orientated techniques are discussed that can be considered in conjunction with the several techniques already described. As stated earlier, there are no hard lines that truly separate different approaches to sea level rise; the categories used here are for conceptual and practical convenience, but the techniques identified under each category blend together when thinking practically about holistic adaptation strategies.

3.3.2.1.3 Hybrid Techniques for Protection

There are a variety of *hybrid techniques* that have been applied to protecting against the incoming sea. Usually these techniques include fixed-positioning armoring techniques combined with other techniques aimed at preventing or mitigating the effects of rising seas. The purpose of these hybrid techniques varies, but they are generally geared at preventing the worst effects of sea level rise on the coastal community while in some cases allowing for maintaining the traditional features of the coastline. For example, in certain areas of the country, access to coastal resources is an important economic activity. Tourism is a vital function of many coastal regions, and much of that tourism is often bound to the use of nearshore resources such as beaches that provide direct access to the ocean. Armoring techniques that allow for the erosion of beach area, such as permanent seawalls, can have a negative impact on tourism revenue and thus is not a preferred policy option. In addition, the coastal landscape is sometimes an important feature of a coastal area for both economic and aesthetic reasons. To maintain the features associated with a particular coastal region, hybrid approaches to sea level rise are being considered and implemented where those features are part of implementing shoreline protection strategies. Below is a summary of some of the hybrid approaches that are either being implemented or considered as part of a strategy to protect against the incoming sea.

Breakwaters are hard structures that are placed in the water some distance from the shoreline and usually parallel to the shoreline. The purpose of the breakwater is twofold. First, it acts as a "speed bump" for incoming waves, causing the waves to break away from shore and thus disperse the energy that accompanies the wave. Second, by dispersing the energy of the wave away from the shoreline, breakwaters serve to mitigate the rates of erosion found along the immediate shoreline. The wave energy that is normally allowed to interact directly with the shore is limited, and thus there is less sediment transport occurring along the shoreline immediately adjacent to the breakwater. So while not being built at the shoreline or inland of a sea–land interface, breakwaters nonetheless serve as an important tool in mitigating the impacts of wave energy. This can be particularly useful as sea level rise brings associated phenomena, including increasing frequency and intensity of storms to a coastal area; breakwaters may be useful in dissipating the increased energy associated with these storms, while still allowing coastal features to migrate inland with sea level rise. Because breakwaters are not built directly on the shore, their impact on shoreline features is limited.*

Jetties and *groins* are two kinds of hybrid techniques that protect against the incoming sea but do not fit directly under a fixed-position armoring method or a flood protection technique (Basco 2003). These hard structures are placed perpendicular to the shoreline, beginning at some point along the shore and moving out some distance into the water. The purpose of these structures is to stop the longshore transport of sediment, which was described in detail in Chapter 2. By preventing sediment from moving parallel along the shore based on current activity, the groin or jetty allows sediment to build up on one side of the structure, while preventing sediment accumulation on the other side of the structure. Strategically placed, the effect of these structures is to nourish a portion of coastal landmass with sediment build-up so as to mitigate the impacts of erosive forces on the target area. Jetties and groins may be used in conjunction with other techniques to provide a stabilizing force for coastal areas that rely on natural current processes to redistribute sand and lessen the need to engage in active management such as beach nourishment projects.

Beach nourishment is a process that adds sediment to the dry beach and intertidal zone areas to reinforce the area against the forces of erosion and other influences stemming from sea level rise. Nourishment can occur naturally through sediment transport, where erosion from wave and current activity in one area leads to the build-up of sediment in another coastal area. The process can also be both passively and actively managed through human intervention. Passive management can occur where a jetty, groin, or breakwater allows for the build-up of sediment in specified areas. Active management can occur where the area to be nourished is identified and sand or other sediment is actively placed at specified sites. As a policy instrument, beach nourishment can be utilized by itself or with other techniques to achieve a particular policy goal in responding to sea level rise. For example, nourishment projects can be combined with breakwaters; the nourishment helps to provide reinforced sediment to the area, while the breakwater helps to mitigate the erosive impact of waves by absorbing most of the wave force prior to the wave reaching the nourished coastal area. Another example is where a combination of nourishment on an ocean-facing side of an island or peninsula is used in conjunction with a seawall on the bay side of the landmass; the nourishment helps to preserve the landmass on the ocean side, while the seawall adds a protective layer from storm events on the bayside. These are just two examples of how several adaptation techniques may be used in combination to accomplish policy goals.

* Breakwaters do tend to slow the transport of sand along the shoreline as they diminish the wave energy reaching the shore. One result of this is that there is less sediment transported to shorelines located adjacent to the breakwater, which means breakwaters may potentially increase erosion rates in areas in close proximity to the area they are protecting.

A prime example of a holistic approach to dealing with sea level rise is the planning and creation of *living shorelines*. A living shoreline is a management technique that prioritizes natural coastal processes for policies that are developed to deal with sea level rise adaptation (National Research Council 2007). Native plants and materials are often used in conjunction with some of the techniques identified above to provide a holistic approach to sea level rise management. For example, native plants are often used to provide stabilizing influences along coastal areas to help mitigate erosion. This planting can be supplemented with groins and breakwaters in order to trap sediment and reduce wave energy. The shoreline area can be nourished with a variety of materials that help prevent the immediate influence of wave activity while biodegrading over time. The benefit of the living shoreline technique is to provide a strategy that allows for natural conditions to dominate in the area as much as possible. For example, if sea level rise were to occur at a rate that exceeded the capacity of wetland vegetation to move inland, then actively transplanting wetland vegetation inland is one option that can be used in conjunction with a few nourishing techniques to slow the rate of sea level rise by increasing the height of the landmass for a short to intermediate time. Again, the priority in a living shoreline ideology is to maintain ecological functions of the coastal area to the extent practicable while also mitigating the impact of sea level rise on the area.

A living shoreline technique does protect against the incoming sea, but its priority of holding back the water is secondary to maintaining the ecological function of the area. In terms of looking at adaptation strategies from the standpoint of prioritizing interests, it may be suggested that most protection strategies against the incoming sea tend to favor humans over environment. The strategies identified below, where the focus is on *adapting* to the incoming sea (rather than protecting against it), can further enhance understanding of how different strategies achieve (or prioritize) different goals. For example, a focus on adapting to the incoming tide while still choosing to remain in the coastal area may allow for consideration of values that are not necessarily available when thinking only about protecting against the incoming sea. A strong, high, well-built seawall may hold back the sea, but choosing that policy technique may protect certain human-based values at the expense of other legitimate values such as the ecological integrity of the area. Understanding the tradeoffs between policy choices can be helpful when trying to choose between the techniques described in this chapter.

3.3.2.2 Adapt to Incoming Sea

The concept of adapting to the incoming sea can be differentiated from protection strategies by the way in which the coastal community views sea level rise. Both kinds of responses are based on the premise that the community will remain along the coastal region. However, the *expectations* of those who stay in a coastal area are different when viewed through a lens of protection versus adaptation. The kinds of techniques described under the protection section above are primarily based on armoring against the sea, that is, preventing the impacts associated with sea level rise. Attached to this notion of prevention is the expectation that coastal landowners will not have to change their behavior patterns over time; the way landowners interact in the coastal zone today will remain the same even with sea level rise because actions will be taken—such as armoring—that prevent the sea from encroaching landward. Thus, even if sea level rise is to occur, the impacts will be limited and, to the extent practicable, never realized because of proactive actions taken to protect shorelines.* Adaptation, on the other hand, suggests a perspective that accepts sea level rise and its impacts as a precondition to determining how best to deal with the phenomenon today *and* in the future. In sum, adaptation strategies generally support the movement of the sea landward and seek to find ways to accommodate human existence in the face of sea level rise. Adaptation strategies, similar to prevention strategies, seek a goal of limiting the impact of sea level rise on the coastal community. However, adaptation differs from protection in that the methods employed are more harmonious with the encroachment of the sea by not seeking to prevent the sea's movement itself, but rather focus more on limiting the impact a moving sea has on existing development. In this way, adaptive approaches can achieve other goals in the face of sea level rise, such as maintaining the natural coastal features of an area to a greater degree than some of the protective strategies identified earlier.

Remember we are discussing techniques that are used to adapt to the incoming sea in conjunction with the choice to remain at the coastal zone (as opposed to retreating from the coastal area—discussed in the next section). So in summary, the approaches described here must consider categorically the best means of allowing humans to remain in a dynamic and changing environment. The goal, then, should be to mitigate the impact of sea level rise on the community in a variety of ways without hard armoring against the tide. Some of the techniques that can be used to accomplish this goal include techniques described above, particularly the use of *dunes* and *nourishment projects*. Both of these techniques are examples where the coastal community is buffering against the impacts of sea level rise without necessarily preventing the rising sea. Such techniques, along with the others summarized

* The idea that protection means preserving human habitation as it stands today does not absolutely correlate to some of the techniques described in the section on protecting against the incoming sea. Hybrid methods, in particular, are techniques that seem to imbue the elements of accepting changes in the coastal area over time. Indeed, the *living shoreline* concept is a technique that seems to embrace the principle of change in that it seeks to preserve elements of the shoreline as sea level rises and moves landward; implied in the goal of maintaining shoreline elements is the acceptance that the shoreline will be changing to include the understanding that certain actions will *not* be taken to prevent the shoreline from moving landward. This reinforces the earlier statement that these strategies fit more honestly on a spectrum that does not easily divide between protection and adaptation. However, the differences between policies aimed at preventing inundation and those adapting to the effects of inundation are important because they influence policy directions.

below, represent a category of approaches where the focus is on changing human interactions to accommodate a rising sea, for example *elevation* of land surfaces and structures, rather than preventing the sea from encroaching inland.

Elevating land and structures encompasses a set of techniques that focuses on mitigating the impacts of sea level rise by raising the height of the land or, where land raising either cannot be done or is insufficient to create a reasonable degree of safety, raising structures so as to keep up with sea level rise (National Research Council 1987). For example, if sea level rise were occurring at a rate of 3 millimeters per year on average, then a project that raised the land around the ocean by 30 millimeters would be sufficient to keep ahead of the current rate of sea level rise for about 10 years. The rationale behind focusing on elevation is to secure the safety of the population along the coast in as reasonable a manner as possible while realizing the benefits of coastal living now and during intermittent periods of sea level rise. Elevating land and structures may be done in conjunction with other methods, including armoring techniques. Indeed, in areas where coastal exposure increases the risk of loss from coastal flooding, local regulations and insurance requirements often mandate that human structures built near the coastline where they are exposed to an increased risk of flooding include design elements that incorporate raising the structure above the anticipated flood level as a condition of receiving a permit to build (public regulation), or as a condition of receiving an insurance policy on the structure to mitigate the risk of loss (private requirement). These requirements of raising the structure above a flood line might exist even in places where armoring has been done to protect against coastal threats.

There are a variety of ways in which elevating the land and structures can occur. For areas that are uninhabited, elevation techniques can often take the form of *filling operations* where nourishment projects occur along the coastal zone (beach) and are continued landward of the beach. A representation of a filling operation to raise the level of the land is shown in Figure 3.4.

The area is graded in a way that protects against sea level rise at all stages of the process. However, coastal wetland filling is something that may be prevented in this process, depending on limitations provided in federal and state laws regarding wetland filling. For developed coastal areas, existing structures can be modified in a few ways. One example is where the land underneath the structure is being elevated. In order to accommodate this process, the existing structure is temporarily lifted using a method that usually employs hydraulic jacks to raise the structure. Fill can be placed beneath the structure itself and the surrounding land. The result

* Historically, wetlands have often been filled as part of dredging and filling operations, where the materials that were dredged in portions of a wet area were often used to help elevate (fill) partially submerged lands in the hope of making those lands more suitable for development purposes. Dredging and filling today in the United States is strictly required by the federal Clean Water Act (for "navigable" waters) and associated state laws that protect wetland resources.

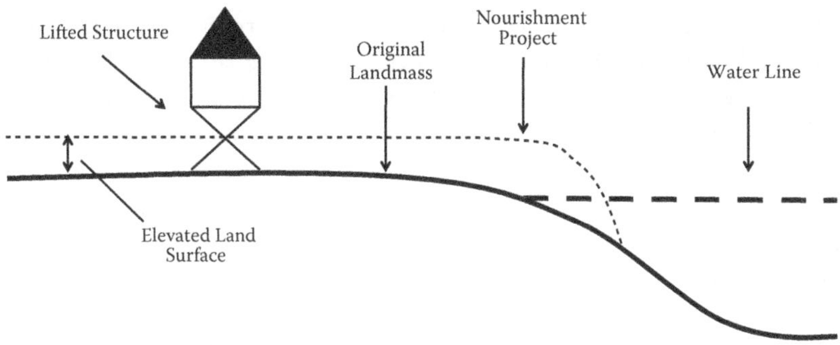

Figure 3.4 Representation of a filling operation to raise the level of land.

is that both the structure and land have been raised by a level appropriate to stem the impacts of sea level rise for the immediate future. Another method focuses on the existing structure itself, employing a methodology to raise the structure but not necessarily the surrounding land. The result is that the structure has been raised in height but the surrounding land has not. This option is primarily meant to prevent intermittent inundation through tidal and storm surges. It has limited use in dealing with sea level rise in areas where the sea will move landward over time. The reason is that the community would be in the situation where their houses would sit atop the sea at some point in the future, with their homes just above the new sea level. Because of this reality, raising existing structures without also raising the land underneath is more of a temporary solution to deal with intermittent flooding than it is a solution to deal with encroaching seas.

Dunes are another means of providing elevation to land masses. They are either naturally created through the dynamics of wind and soil along coastal areas, or they can be physically created through human intervention. Dunes serve a variety of purposes (noted earlier) that include habitat for coastal species, soil stabilization, a source of nearshore nourishment, and also as a barrier to coastal forces impacting areas landward of the dune system. As an adaptive strategy, the services provided by dunes can be utilized to help coastal communities adapt to sea level rise. The dune system itself can mitigate near-term sea level rise by providing both a barrier and source of nourishment to combat the early march of the sea inland. Also, dunes can mitigate the impact of sea-borne storms where the forces of those storms would be moving further inland in step with the ocean's rise. Where dune systems do not exist, the development of human-made dunes can provide some or all of these benefits as a means of adaptation. Of course, most of these measures alone will provide temporary benefits but will not solve large-scale sea level rise unless the dunes are large enough (and possibly maintained) to ensure a substantial barrier that limits sea level ingress on land.

Dredging along the coastal area is another way to modify coastal land to adapt to the incoming seas (National Research Council 1987). Rather than raising the landmass through elevation projects, the capacity of certain parts of the land can be increased to absorb the incoming water as it moves landward. Dredging is a process where low-lying areas of land (generally estuary areas where the land is partially wet) are dug out so that the area can accept a larger volume of water. Sea level rise may be mitigated through this process in certain regions where there is a capacity of low-lying areas to accept additional water from the ocean in a way that mitigates the rise of the water into neighboring coastal communities. Of course, the extent to which any dredging operation might meaningfully lessen the impact of incoming tides depends on the extent of the dredging as well as the associated geophysical characteristics of the region. The costs borne through these projects may be justified in regions where dense coastal population centers are located and the expense of a large-scale dredging project to manage sea level rise is justified because of the potential population impacted and assets at risk in the area. The same issues regarding wetland protection apply to dredging operations here as they did to fill operations discussed above. Any dredging of a wetland within federal jurisdiction requires a permit under the Federal Clean Water Act, and there is some question as to the kinds of activities that might be allowed in anticipation of sea level rise where the harm to the wetland system is known but the corresponding risks to the coastal community are unknown. Under regulatory regimes where wetland protection is prioritized, it may be difficult to justify a proactive approach that incorporates dredging as a means of protecting sea level rise that has yet to be realized.

In addition to dredging and filling operations, the other main way to adapt to sea level rise while remaining in the area is to simply apply a *wait-and-see* approach. As waters rise over time, the community may respond to incremental rises using some of the techniques described above. For example, if it is feasible to raise the land in the most at-risk areas, then this may be a superior option. Alternatively, a mix of nourishment projects that focuses on dune maintenance/creation and also small-scale dredging operations may provide a near-term solution to what is being observed. The main benefit behind a wait-and-see approach is that no pre-commitment of resources is required; resources are only committed once the problem is fully realized and understood. Of course, it may be that once the problem is fully realized, the decision to remain in the coastal area may have to be revisited. For example, sea level rise may begin to occur in certain regions at a higher-than-expected rate and thus have impacts that diminish the original assumptions about staying in the coastal zone. Under these circumstances, the decision to stay may be exchanged for a decision to *retreat* from the coast. The choice to retreat can take many forms and includes a variety of considerations. However, from a policy standpoint, the choice to retreat presents a different set of options in terms of the kinds of resources that may be committed toward mitigating the impacts of sea level rise, and also the relative value assigned toward different resources that may be impacted by the choice to either stay or retreat.

3.3.3 Retreat

Retreat in the context of adapting to sea level rise is meant to refer to a *planned process* incorporating policy techniques supporting a proposition that the current coastal zone will no longer be habitable (or the costs of staying will exceed the benefits) some time in the future due to the impacts caused by sea level rise (Titus 1990).* Planning can occur in a number of ways, including purposefully choosing not to armor areas that are in obvious need of shoreline protection and therefore using a planned retreat strategy as a means of dealing with the dangers posed to the shoreline as an alternative to armoring. A planned retreat policy may also be used in areas that are not highly sensitive to the dangers of sea level rise, but are implemented to decrease the costs associated with a retreat should the condition be required in the future (U.S. Climate Change Science Program 2009, p. 87–104). In both cases, retreat as a policy goal overrides protection options as a way of dealing with sea level rise. From there the choice of what kinds of techniques to best employ retreat as a policy goal becomes the defining characteristic of policy development.

Retreat also differs from protection as a major policy directive in terms of the *lead-time* necessary to implement a retreat policy. Implementing shore protection techniques can occur over a relatively short period of time; most of the kinds of projects related to shore protection (armoring, etc.) take months to years to implement. Retreat, on the other hand, is a more comprehensive policy approach that requires a good deal more planning, often requiring years to decades to fully implement and realize (Titus 1998; Randall 2003). Consider, for example, that retreat may include a variety of steps that are both protective of existing structures and also precautionary in terms of future development. A policy framework that carries out retreat goals may begin by identifying the most sensitive areas along coastal regions that may reasonably be impacted by sea level rise in the future. Once identified, this area may be placed under a special zoning district that identifies and highlights the area for management purposes. The first step of policy implementation may include physically relocating existing structures that lie close to the shore in the greatest zone of danger so that this immediate threat is reduced. A second step may be developing special building requirements in this zone that may include preventing building in the most hazardous portions of the zone, and possibly also including certain building standards and setback requirements for new development in less hazardous areas of this zone. Also, "triggering events" such as sea level rise of a certain magnitude may be planned into the management scheme requiring further

* Retreat may also be *unplanned* as is often the case for coastal areas that have been inundated by a storm event. Although unplanned retreats can include some of the same policy techniques described in a planned retreat scenario, one major difference is that a planned retreat takes proactive actions that incorporate retreat strategies prior to the happening of the event that triggers the retreat. In a sea level rise adaptation scenario, the focus is on planning for the sea level rise rather than responding to it. Thus, the focus here is on a planned retreat from the coast that incorporates proactive planning into the decision-making process.

actions such as a rezoning of the area to reflect new dangers; the rezoning itself can then trigger additional requirements such as relocating structures in new hazardous zones. Collectively, retreat planning is more comprehensive and generally more proactive than staying and protecting the shoreline. As a result, retreat planning takes more time to develop and implement than shoreline protection strategies, although shoreline protection can also be planned in advance in a more comprehensive manner.*

The choice to *retreat* from the shore rather than to *stay* also provides a different set of policy options for consideration; the policy options are different in terms of the options identified, where the options for retreat often reflect preferences for values that extend beyond human well-being as a priority in the coastal zone. Considering resource expenditures, the choice to stay at the coast can provide unique up-front costs that otherwise may not be necessary when retreat is chosen as the accepted goal in coastal management. Often, substantial resources must be committed when the major policy goal focuses on staying at the shore, either through armoring or other techniques, with no guaranteed assurance that they will succeed over time. The choice to stay includes some commitment by government to protect against both real threats and also potential threats. This may include instances where resources are expended well in advance of sea level rise to assuage the local population who may demand such resource expenditures to reinforce the idea that there will be protections against future sea level rise.†

Retreat focuses on managing *human expectations* rather than on managing the *forces of nature,* as is often the goal associated with staying in a coastal region (Randall 2003; U.S. Climate Change Science Program 2009, p. 87–104). In a retreat policy

---

* As noted elsewhere in this chapter, the choice to *stay* at the shoreline and protect or adapt to the incoming sea can be based on a more comprehensive planning strategy that incorporates elements of armoring, adaptation, and even ultimately retreat, depending on the circumstances. However, basic shoreline protection strategies are the kinds of policy instruments that can be implemented quickly as long as resources can be allocated to the task in a timely fashion. A planned retreat is a kind of policy framework that is more proactive and comprehensive in its formulation and development. It focuses more on changing practices and behavior patterns over time so that a greater degree of resiliency is created in the community, making it easier to respond to sea level rise in the future.

† Recall the earlier discussion about the relative benefits and costs where the uncertainty of sea level rise justified waiting to expend resources until the uncertainty of the rise was more likely known. This is true in a rational analysis of sea level rise adaptation where expenditures are only done when necessary and to the extent required to alleviate the risk. However, as stated in the earlier section, a commitment of resources to an area like developing in a coastal zone often includes assurances that help mitigate risks even when the probability of the risk is not known. Thus, public entities sometimes may be pushed to expend resources today to protect against an unknown risk tomorrow where the expenditure of resources helps to alleviate risk perception (private parties are more likely to develop along a coastal region that is pre-armored against a potential rising tide because the armoring represents a greater likelihood that the area will be protected in the future should an event like sea level rise occur, regardless of the probability).

framework, public and private entities internalize the reality that leaving the coastal zone is a possibility in the future. In this way, people tend to manage their decision making, including how they expend their resources, in a way that incorporates this potential. Contrasting retreat with a philosophy of staying at the coast, the kinds of policies that develop can be quite different as the management of human expectations takes a central role in policy planning, whereas the management of nature tends to highlight technological capacity over the elements without acknowledging the stark possibilities of how sea level rise may impact those expectations over time. This lens of "expectations" is a useful way of thinking about how policy develops around underlying presumptions. It may also give some clues as to the kinds of techniques made available in a retreat setting versus a stay setting. As such, it is worth taking some time now to explore this distinction in greater detail before reviewing the different kinds of techniques used in a retreat policy framework.

3.3.3.1 Community Expectations: Retreat versus Stay

To begin, consider Figure 3.5, which shows a conceptualization between a policy geared toward retreating and one aimed at staying in the coastal zone.

As noted above, one way to conceptualize differences between staying and retreating is to focus on the expectations created by each policy direction. More particularly, the differences between expectations under each approach can be

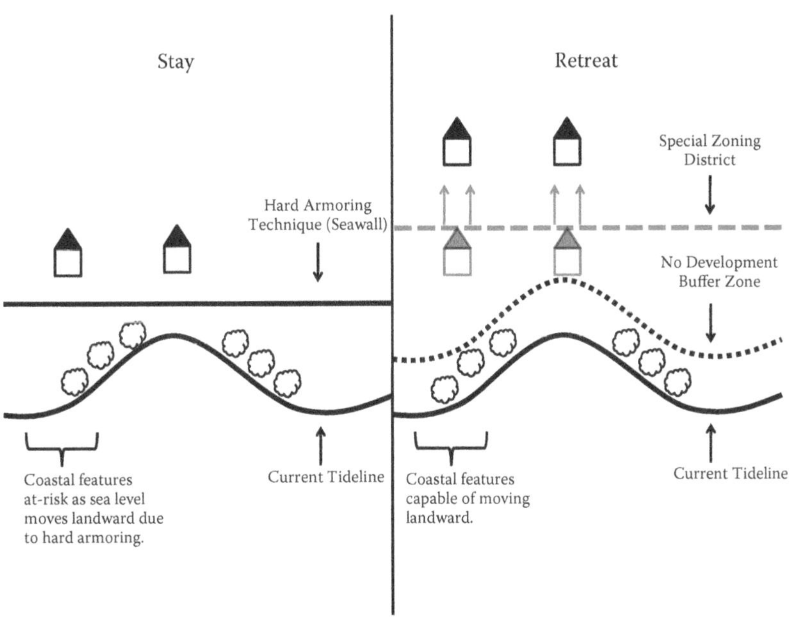

Figure 3.5 Conceptualization of a retreat and stay policy in a coastal area.

analyzed through a framework that focuses on a net benefit determination where uncertainty is an important factor in determining net benefit.* If the choices of staying and retreating were viewed on a continuum including different degrees of staying and retreating, then a greater degree of resolution between these options might be available in order to understand the major differences. For example, let us assume there is an extreme version of staying as a policy goal. In this version, the community is committed to staying in the coastal area, regardless of the extent of sea level rise. Thus, there are no contingency plans for leaving the area under any policy considerations; there are only policy options that reinforce staying and the kinds of coastal management techniques employed only reinforce staying as an option. At the other end of the spectrum is an extreme of retreat, where the central theme of the policy goal is to leave the area due to the potential of future sea level rise alone. In other words, the community is set on leaving the area even without any signs of sea level rise being realized; under this policy, the community will not wait to be pushed out by sea level rise, but rather will leave well in advance of sea level rise. In the middle of these two extremes are variations on each theme, depending on where the goal falls within the spectrum. Those goals that exist closer to the middle of the spectrum equally balance the staying and retreating options. Goals that lean more toward the staying extreme employ more techniques of staying than techniques of retreating, and goals that lean more toward retreat employ proportionately more retreat techniques than stay techniques.

So what are the net benefits of staying versus retreating, and how can these different net benefits be viewed in relation to a spectrum where several techniques are employed? Answers depend in large part on the expectations of the community, which are partially defined by the early planning undertaken by the public entity charged with developing coastal management policy in the area. Consider a community where retreat has been defined as a management option for long-term planning that is balanced against short- and medium-term options that include staying. In essence, consider a policy framework where the focus is on waiting to see the extent of sea level rise in the area, but incorporates elements of both staying in the near- to mid-term with elements of retreating should certain conditions come to pass. This policy framework might include a variety of soft-armoring options that help to reinforce the status quo in the coastal community; existing development is supported through techniques such as beach nourishment, offshore stabilization (breakwaters), and living shorelines that aim at mitigating the near-term impacts of sea level rise. These activities may be coupled with policies that limit coastal development in sensitive areas near the coastline. For example, there may be a ban on development in particularly sensitive areas where present erosion rates are high, thus giving a signal that these areas would be particularly susceptible to future sea level rise. Alternatively, building standards may be different within a defined coastal area where greater precautions are taken to ensure new buildings

* This framework is discussed in greater detail in Section 3.2.

are resilient to near- and mid-term sea level rise and the associated storms that may accompany such rise. The building standards might require the structure to be elevated off the ground and/or designed in such a way that allows the structure to be easily moved landward from the area. There may also be policies that are geared toward a minimum setback from the sea that reflects currently observed erosion rates with some additional precaution for sea level rise. This setback requirement would thus be based on ensuring a greater probability that structures developed in the area would be capable of habitation even with a moderate degree of sea level rise for a reasonable period of time.*

Focusing on the minimum setback requirement based on erosion rates as a technique for managing expectations helps place the idea of net benefits and uncertainty into context when it comes to distinguishing between sea level rise adaptation policies aimed at staying in the coastal area versus polices aimed at retreating from the coastal area. Consider a minimum setback requirement that contains the following elements that are internalized by the coastal property owner as conditions for developing within the coastal area:

- The setback requirement is based on an understanding that the area is dynamic and sea level rise may someday inundate the area.†
- The setback is based on a best estimate that suggests the property will be viable for habitation for a period of time (with no guarantee).
- The setback tells the property owner that attempts to hold back the sea will not be a viable option for defending the property in the future.

---

* Setback requirements are often based on considerations that are human based (health and safety considerations). Thus, a setback requirement from major infrastructure or property lines of 30 feet is usually not based on a consideration of the dynamic forces involved in coastal areas. A setback requirement that is based on internalizing the likelihood of sea level rise in the area can be helpful in ensuring that new structures are built with a reasonable expectation to survive long enough so the investment in development results in a net benefit (where the benefits of the structure's existence exceed the costs of development). Thus, a setback requirement that provides a likelihood of the structure being habitable for 30 to 50 years may be sufficient to ensure the reasonable use of the structure and justify development. Note that this kind of setback requirement implies no hard armoring will be allowed to protect the structure against the rising sea. The expectation of the landowner in developing the structure is that at some point in the future, the building will have to either be moved or abandoned should sea level rise require this result.

† The setback requirement itself is generally part of a land-use management scheme that internalizes the potential for sea level rise by shifting the risk from the public entity to the private landowner through conditioning development of the property based on factors that realize the potential impact of sea level rise. Language in the setback requirement can include provisions that indicate the reason for the setback, how it is determined, and what happens should water rise to a level that exceeds the setback requirement.

- The setback requirement is a condition that encapsulates the risks of sea level rise into the decision-making process of the private landowner; the very existence of a setback requirement based on likelihood of sea level rise (via erosion) helps the private party internalize the potential costs of developing *before* the development occurs.

The very existence of the policy instrument of a development setback requirement that is based on incorporating sea level rise considerations into the setback itself creates different expectations between the private landowner and the public entity. For example, the setback allows for costs to be incurred in a manner that most efficiently provides for net benefits to be realized even under a strong level of uncertainty regarding sea level rise. If a private landowner decides to develop in the coastal zone under the conditions of a setback requirement that internalizes sea level rise risks (even with uncertainty), then there is a greater chance the private landowner is identifying net benefits where the overall benefits of building the structure exceed the costs of development *even under the constraints of the limited life cycle of the coastal project due to the setback requirement.** This means the project has been deemed valuable, regardless of sea level rise in the future; if sea level rise occurs and the life of the project is limited, then the value has still been realized because the decision to have a limited time in which the development was viable due to the possibility of sea level rise was made prior to development. If sea level rise does not occur, or occurs more slowly than anticipated, then the development project maintains benefits *beyond* what was originally expected. In either situation, the policy allows for a benefit-cost analysis that internalizes risks *before* the project is undertaken, thus allowing net benefits to be realized even under conditions of uncertainty.

Contrast the setback requirement example above with a public project that engages in hard armoring of the shoreline as a precaution to future sea level rise. Recall that armoring techniques (e.g., building a seawall or revetment) are an example of policy techniques generally geared toward staying in the coastal area. An armoring technique may be done as a precaution to future sea level rise and associated conditions such as increased coastal storm intensity; it may be done in

* It is certainly possible that a setback requirement based on rates of erosion may result in certain parcels of coastal property being deemed "undevelopable" because there is simply not enough land on which to both meet the minimum required setback and develop the property (e.g., a parcel of private land that is 50 feet in total from the shore cannot incur a 50-foot setback requirement because the setback takes up the entire parcel's width – there is no space left to build the structure). If this new setback requirement (based on erosion rates or future sea level rise determinations) exceeds the previous setback requirement, then there may be a technical problem because the new requirement has essentially made previously developable land no longer developable. The new setback may be seen as a *regulatory taking* of the right to develop on the property, and the fact that it is based on an uncertain potential of sea level rise can further complicate whether or not a taking is found. These kinds of complications are discussed in detail in Chapter 4 on legal considerations related to sea level rise adaptation strategies.

response to a recent coastal storm that caused temporary inundation of the coastal area; or it may be done for a combination of these reasons. Whatever the reason(s) for the project, it is an example where resources are being fully expended today (the building of the seawall) with no guarantee that those expenditures (costs) will achieve benefits today or in the future that exceed the costs.* Thus, the net benefit of the activity remains uncertain. However, the costs of the activity can be calculated in *both* the presence and absence of future sea level rise. If future sea level rise occurs and exceeds the capacity of the seawall as a protective feature, then reliance on the wall as an excuse to stay will have substantial costs attached to the project that must be weighed against the benefits achieved in the time period between building the wall and the point of breach by the rising sea. In the absence of sea level rise that threatens to breach the wall, the cost of building the wall must be balanced against the lack of need for the wall as a protective measure. In addition, the existence of the wall may have diminished other values of the coastal area, such as ecological values, and the harm to these values should be considered in the overall costs to the area.†

Another consideration is the effect that the development of a seawall, or other protective technique, has on those living in the coastal community in relation to their individual and collective expectations about remaining in the coastal area. A publicly funded armoring project reinforces the concept that the area is being made safe for habitation today and tomorrow. Thus, it is not unreasonable for those living in the coastal area to expect that any commitment of their resources today will be able to be realized long term and regardless of the background environmental conditions that come to pass. This not only can reinforce the choice to develop in a coastal area, but it can also make it much more difficult for government to change this expectation in the future. So, in cases where sea level rise becomes a serious issue for a coastal area in the future, the early commitment to build protective structures helps create *legacy issues* for government, which means substantial resources may be spent in the future to try and prevent the impacts of climate change even in situations where it is clear that a net benefit (the return on the investment) is negative. In essence, the past commitment of establishing a protective strategy that reasonably creates the expectation of staying in a coastal region creates an ongoing commitment from the public entity to do "whatever it takes" to keep that promise. This is not an ideal situation for policy development: It does not allow a rational set of choices to be considered while maintaining the greatest degree of freedom in

* As explained previously, there are immediate indirect benefits in the creation of the seawall even if the wall is never actually utilized to prevent the impacts of a storm (storm surge) or prevent sea level rise from moving landward. The creation of the wall acts to create a kind of guarantee against future harm that helps make development decisions and the commitment of resources today more clear. This risk-shifting exercise may also lower insurance rates in the area and thus collectively provide incentives to develop in the coastal region today where those decisions might not be made as easily if the seawall did not exist.

† Ecological considerations are discussed in greater detail in Section 3.4.3.

balancing various interests between choices. Rather, it limits policy choices to those that only allow protection techniques as a viable course of action under a goal of staying along the coastline, until of course the conditions become such that there is no alternative but to retreat.

The expectations created between policies that distinguish staying versus retreating as a central theme are substantial. The goal of staying at its extreme is really a commitment to protect against a future event that may or may not occur. As a result, there are expenses incurred that often do not allow for providing greater freedom between policy directions as more information comes to light, thus diminishing the amount of uncertainty about the future extent of sea level rise. Retreat options provide, programmatically, more freedom because they internalize the worst-case scenario of the future event (sea level rise that forces the community to abandon the area) and develop policies that assume this situation. While some costs may be created up-front because of this assumption (setback requirements that push building further back from the shore than may ultimately be proven necessary is one example), most of the costs are only realized upon the occurrence of the future event. If the future event does not occur (or occurs to a lesser degree than anticipated), then the costs are never realized, in that the community remains in the area prepared to retreat should sea level rise occur to a point where this is necessary. Highlighting this difference in expectations between these two main approaches now is one way of establishing a foundation upon which the policy techniques available for retreat may be understood.

3.3.3.2 Policy Techniques for Retreating from the Shore

Specific policy techniques geared toward retreating from the sea tend to follow a greater degree of planning from a proactive standpoint; the techniques tend to focus on proactive and comprehensive procedures that both mitigate the impacts of near-term sea level rise and also plan for easier responses and transition to future sea level rise. Some of the techniques provided in a planned retreat orientation are specific—for example, relocating existing structures that are near the tideline and thus most at risk for inundation or storm damage associated with sea level rise. Other techniques create programs that offer ways of compensating private landowners who own sensitive lands by providing funds for the purchase of those lands. Most programmatic of the policy techniques are the regulations that help manage the expectations of those living in the coastal area. Some regulations are geared toward changing the land-use (zoning) designations of the area to reflect the increased risk associated with sea level rise. Other regulations seek to rely on the police powers of the state to establish development restrictions and building standards that incorporate the capacity to deal with sea level rise in the future. A summary of the major techniques employed in retreat strategies is outlined below. Consider how each strategy can be combined with other strategies to provide a comprehensive approach to dealing with sea level rise from the perspective of retreat, particularly how each strategy works to create unique expectations in coastal land users about the impermanent nature of the area, and thus the likely need to abandon the area in the future if conditions dictate such a response.

The kinds of techniques available for planned retreat include a mix of *mandatory* and *participatory* programs. The term *mandatory* used here is meant to refer to programs that utilize a regulatory power of government in relation to land use. So land-use planning, zoning, and other police powers related to the regulation of land represent the kinds of techniques employed under a mandatory program. Private land and private landowners that fall within the scope of the mandatory regulation have little freedom in complying with the regulations once established; in essence, these regulations are generally compulsory, and private parties must seek redress from their application either through judicial avenues, if available, or through the political process. *Participatory* programs are programs that provide for goals related to a planned retreat but they are not compulsory activities per se; in participatory programs, the choice to engage in the program is often left to the private landowner. There is no compulsory nature to these programs, and private landowners can often opt out of such programs. In addition, both public and private organizations may participate in offering participatory programs.

One of the specific strategies employed in a retreat policy paradigm is the *relocation* of structures out of hazardous coastal areas by moving them inland. Relocation is, in essence, the decision to move an existing structure further inland from its current location (U.S. Climate Change Science Program 2009, p. 87–104). As a tool, relocation can offer a means by which existing structures can be managed in a way that allows for the option of moving them inland based on a variety of criteria. For example, a public program may identify a critical zone within a coastal area based on observed and predicted sea level rise. Buildings located in this zone may be legally required to be moved landward (*mandatory*), or otherwise there may be no legal requirement to move the building but rather a program that fully pays or defrays the cost of moving the building landward (*participatory*). There are a number of issues associated with relocation. For example, a requirement to move the property landward can be frustrated by a lack of available land for the property owner. In addition, the basis for such a requirement can be challenged as an unconstitutional taking of private property where the regulation orders the moving of the property but provides no compensation. Thus, voluntary relocation options are often preferred because they provide an incentive to move property landward with the private landowner's consent and therefore without a legal challenge to the government action. Other complementary programs, such as increasing insurance premiums to reflect the increasing risk of being located in a hazardous zone, may also aid in the success of relocating properties landward of a coastal area.

Acquisition of coastal land is another technique that is used to provide a planned retreat from the shoreline. The idea behind the acquisition is to transfer title of the property from a private landowner to a public entity or nonprofit organization for the purpose of "retiring" the land from development or other human uses. The land remains in an undeveloped state and then becomes part of a buffer for the coastal area to help mitigate the impacts of sea level rise while also ensuring that development occurs sufficiently inland to protect that development from future sea level rise. In theory, such a system could operate continually, allowing for the purchase and retirement of land closest to the water on an ongoing basis. This can be done either in anticipation of sea level rise to establish a sufficient buffer between developed land and anticipated sea level rise, or it can be done incrementally, where new purchases are made as actual sea level rise triggers new acquisitions based on the location of buildings in relation to a moving buffer zone.

There are several kinds of acquisition programs that may be implemented to support planned retreat policies. Reactive programs include acquisitions through *buyouts,* where public financing is used to purchase coastal land that has been impacted by a coastal hazard (Lewis 2012). Often, the triggering event of the hazard makes a private coastal landowner less willing to continue with the risks of living in the area. The buyout is offered to provide some compensation to the private party while allowing the public entity to remove a proven coastal risk by ensuring that the land purchased is removed from future development. The more proactive approach includes a programmatic scheme to identify and actively pursue parcels of coastal land that are located in sensitive areas. The goal is to obtain title to these coastal lands so they can be removed from development for human safety purposes and also so that the natural features of the coastal zone can move inland uninterrupted by human development. The acquisition of title can be done through voluntary measures where the private party gives title to the land or places a voluntary restriction on the land that prevents development,* or it can be done through involuntary measures such as eminent domain proceedings or development regulations that require certain at-risk areas to remain undeveloped as a condition of permitting a development project on property in a less risky area. Both public and private organizations engage in these proactive kinds of acquisition programs. From a policy standpoint, the process of acquiring lands can be expensive if done through eminent domain procedures (particularly where the land is being valued as a developable piece of coastal property). As such, regulations that focus on permitting conditions and voluntary measures may be less expensive methods for acquiring coastal lands because they require less capital investment in the property by the public.

Regulatory restrictions form a category of techniques often employed as part of a more programmatic scheme to establish a planned retreat ideology in the coastal

* One of the most regularly employed types of voluntary restriction utilized on private land is the *conservation easement.* The easement is a restriction usually placed into the title of the land by the present owner of that land. The goal of the easement is to create enforceable rights against the development of certain parts of the land, effectively removing certain portions of the land from development. These easements may be purchased by the holder of the easement (public or private entity), and public "purchases" often include or are wholly accomplished by a reduction in property tax assessment on the land placed into the development restriction.

community (McGuire and Hill 2012). The techniques employed here are founded in public regulations that prevent the land from being used in a certain fashion, or otherwise condition the use of portions of private land on the condition of not using other portions of the same land. The use of regulatory restrictions allows for limited development in coastal areas deemed most at risk for sea level rise, ultimately pushing any development of the area further upland from the coastline. This accomplishes the goal of protecting human habitation from sea level rise by moving habitation inland and away from the coastal threats, while also accomplishing the goal of allowing coastal features to migrate in step with sea level rise.

The kinds of regulatory restrictions that exist along the coastal zone generally focus on conditions placed on development. As noted above, acquiring public title to coastal land may be accomplished by conditioning development on the granting of public title to certain portions of private land, but the "title" granted is usually based on a development restriction that allows development on part of the land in consideration of preventing development on other parts of the land. Development restrictions that do not focus on acquiring an interest in the land generally fall into a few categories: *setback requirements*, *flood hazard regulations*, and *rolling easements*. Each of these techniques is often applied as either a part of the permitting process for development itself (meaning that the restriction is not triggered unless/until an application to develop is made), or the technique is made part of a general planning strategy and thus applies uniformly to all property along a coastal area regardless of whether or not the landowner is seeking development.

Setback requirements are regulations that mandate a development can only occur beyond a minimum distance from a defined boundary, say a road, for example (Kim 2012).* Setback requirements for coastal development can be based on annual erosion rates or other criteria that allow some assurance that the development will not be in jeopardy from sea level rise for a reasonable period of time. For example, a property located along the shoreline might have a setback requirement based on an established "erosion rate" to ensure there is a reasonable probability the development will last for the period of time indicated. Many times, the erosion rates are based on the life cycle that follows traditional mortgage financing options. A traditional 30-year mortgage for a single-family residence might establish the basis for the erosion setback requirement; the setback will be based on an amount of distance that provides some reasonable probability that erosion rates will not impact the home for at least 30 years. The manner in which the erosion rate is established to create the setback minimum can vary but is usually based on a combination of historical rates of erosion documented, current erosion rates, and future predictions of sea level rise for the area. If an erosion rate was established at 1 foot per year based on the factors identified above, and further there was a 30-year window established for the useful minimum life of the property, then the setback would be

* If the setback requirement was 30 feet from a road, then a property that abuts the road must have at least 30 feet between the road and the closest part of the structure.

30 feet (that is, 1 foot/year (times 9) 30 years). The actual setback might be greater than 30 feet in order to accommodate accelerated rates of erosion or increased exposure to tides and storms based on the unique physical characteristics of the particular coastal zone where the property is located.

Flood hazard regulations may be employed to prevent development in certain low-lying areas near the coastline that are subject to flooding and thus possible inundation through sea level rise (Titus 1998). The basis for a flood hazard assessment is elevation of the area and not necessarily proximity to the shoreline. The process of enacting such restrictions usually includes identifying flood-prone areas through the creation and/or interpretation of floodplain maps. Once identified, these areas are limited in terms of development, sometimes outright banning the use of these areas for building purposes. As a policy tool, preventing construction in floodplains can aid in creating important buffer zones within a watershed area that extends beyond the immediate coastline because these areas will be subject to inundation as sea level rises and the ocean water finds the nearest and lowest spots as it migrates inland. In addition, preventing development in these areas provides open space for the migration of wetlands and other coastal features to allow for maintaining the ecological integrity of the area.

Rolling easements provide another important programmatic tool used in planning and managing the retreat of a coastal area. A rolling easement is a regulatory mechanism that delineates the interests in land between public and private parties. In essence, a rolling easement, triggered by seas moving landward, transitions ownership of once privately held uplands into publicly held submerged land. In addition, the upland immediately adjacent to the sea may also change its character, providing rights of access and use on behalf of the public and thus supplanting private right interests that may have existed on the dry land prior to the movement of the sea inland (Titus 1998). The effect of this transition of property rights allows the characteristics of the coastal zone to move landward—both the physical characteristics of the coastal zone itself and the legal characteristics related to property rights. Newly submerged land moves from private to public ownership, and the dry land immediately adjacent to the water line may or may not contain public rights of access, depending on the state's laws related to public rights in the nearshore.*

From a policy standpoint, rolling easements may be particularly useful in managing community expectations in the direction of planned retreat. For one, rolling easements prevent private armoring of the shoreline and thus help to reinforce the idea that the retreat is a possibility in the future. In addition, the concept that

* Submerged lands are generally owned by the public and managed under the *public trust doctrine* by the respective government for the benefit of the public. In addition, the public may maintain rights of access and use in the nearshore dry land immediately adjacent to the water based on public trust rights established by the state under principles of background law as defined in that state. Ownership of submerged lands as well as public trust rights and obligations are discussed in greater detail in Chapter 4.

property rights will be adjusted as the tide moves landward creates a further set of expectations that coastal property is impermanent and therefore retreat is a likely outcome should sea level rise erode the private landowner's property rights over time. Finally, a rolling easement regulation allows for similar objectives as acquisition projects or setback requirements; the buffer intended through such projects is also available through a rolling easement as armoring and other techniques to hold back the tide are not allowed. Thus, the goal of ensuring that the buffer exists between development and the tideline is met; however, the rolling easement is more comprehensive in scope and cheaper than implementing purchasing programs or eminent domain proceedings. In sum, rolling easements provide a mechanism to control land-use patterns in a way that also controls landowner expectations without substantial government investment in the process. Private landowners may also be more likely to plan for retreat under a rolling easement restriction because the regulation assumes the eventual loss of property (including legal rights) before sea level rise occurs, and this helps private landowners internalize this future loss when making current decisions about investments in their coastal property. If sea level rise does not disrupt the private use of land during the time it is owned by a private landowner, then full value for that owner has been realized. However, if it does occur during the landowner's tenure, then the landowner will have an understanding of the loss before it occurs and can plan accordingly for that loss.*

Zoning regulations are another technique employed to plan for a future retreat that may be necessary due to sea level rise. Zoning as described here is meant to apply to general zoning restrictions that are geographically isolated to coastal areas with the intended purpose of limiting the number of people residing in those coastal areas (McGuire and Hill 2012). Two types of zoning regulations that have often been utilized to restrict the number of people are *density restrictions* and *size limitations.* Density restrictions allow development to occur in a coastal area but limit the amount of development that occurs. The motivation behind the restriction can be aimed at protecting against tidal storms and other dangers that linger in coastal regions. Often, the purpose of existing density restrictions is to limit the

* The ability to plan for the loss brings in other elements discussed earlier in this chapter. For example, landowners who understand that their property is subject to a rolling easement may include design elements in the construction of their homes that allows the home to be easily removed from the property; this decision to engage in such design standards may be voluntary or compulsory (building code regulation under adaptation strategies). Alternatively, landowners may wish to build their structures further back on their property in anticipation of a rising tide to ensure they maximize the use of their property for the longest period of time possible. Also, a landowner might choose to construct a simplified, lower-cost structure to mitigate the potential loss of that property if sea level rise occurs. There are a number of ways in which the regulatory setting can alter landowners' expectations about the permanence of their property and thus allow for behavioral changes that internalize future risks today. When those behavior changes result in choices that reinforce a planned retreat strategy, the policy goals are being met through individual decisions.

amount of runoff and the potential for pollution in nearby coastal estuaries and bays. However, as a policy tool, density restrictions can aid in planned retreats by ensuring that the coastal area has a limited number of residences that may be impacted by future sea level rise.

Size limitations are the other form of zoning regulation. Like density restrictions, size limitations allow development in the coastal region, but a limit is placed on the structure's size. The rationale is that smaller structures require less use of surrounding resources. For example, in areas where septic systems dominate and contamination of groundwater is a concern, a size restriction on the footprint of the home can help ensure a minimum level of protection (along with density restrictions). From a retreat standpoint, smaller-sized homes are easier to move; combining a smaller footprint with design standards that make the homes easy to break down is one way of building the capacity for a planned retreat in the event sea level rise occurs to the point where retreat is necessary. Taken together, density restrictions and limitations on the size of particular properties located in the coastal zone are two ways of planning for a retreat from sea level rise that reduce the burden of retreat while still allowing the coastal property to be put to some use today.

Retreating from the shoreline is a policy goal that is fundamentally different from the choice to stay. Staying implies limiting choices about how to deal with sea level rise simply because the choice to remain at the coastal zone has been made prior to experiencing the impacts of sea level rise. In situations where sea level rise occurs to a point where retreat would be a viable option, the choice to stay limits policy options to some mix of protecting against the incoming sea and adapting behavior patterns as one remains in the coastal zone. Of course, the extent of sea level rise may result in abandoning a choice to stay at the shoreline, but in such a case the policy options available are limited to abandonment and evacuation to two policy choices that are not the kind of "planned retreat" being described in this section.

The conscious choice to retreat prior to sea level rise and thus forcing a community to abandon the coastal area provides more freedom in the kinds of policy actions undertaken both today and in the future. The example techniques identified in this section supporting a planned retreat focus on a mix of voluntary and compulsory mechanisms to manage *human expectations* in the coastal zone. The idea that land-use planning mechanisms help to increase *capacity* to leave a coastal area well before sea level rise or associated coastal storms mandate such an action ensures that the potential of sea level rise has been internalized into the behavioral fabric of the coastal community well in advance of any actual sea level rise that might trigger the necessary retreat. This internalization is important as it establishes the foundation upon which community expectations are defined. For example, an ongoing deployment of planned retreat policies allows a degree of retreat to already be undertaken in the coastal community; the mix of mandatory regulations (setbacks, easements, density and intensity restrictions) and voluntary policies (purchasing programs, conservation easements) provides for the establishment of a buffer between human habitation and the coastline. Thus, the coastal community already has built-in resiliency toward sea level rise. In addition, building standards and other zoning options can increase the capacity for structures to both withstand the increased volatility that may accompany sea level rise, as well as provide a greater capacity to move existing structures inland.

One of the more compelling differences between retreat and stay strategies is the level of community expectation placed on the government should sea level rise occur. An overall policy stance to stay in the coastal area creates the expectation that government will do whatever is necessary to meet that objective; the coastal community will expect government investment to meet (and likely exceed) the private investments undertaken based on a belief that staying in the coastal area is the long-term goal. Hard armoring public projects (such as building seawalls) helps to both create and reinforce a community expectation that government will engage in whatever efforts are necessary to protect the community against the incoming tide. This expectation can reinforce government actions that further reinforce the community expectation of staying in the coastal zone. This kind of feedback cycle can collectively provide unreasonable expectations (and outcomes) should sea level rise occur to a point that abandonment of the area is required. Contrast this setting with the implementation of policies that plan for future sea level rise today without substantially compromising the ability to enjoy the values and attributes associated with living along the coast. By internalizing the potential of future sea level rise through a strategy of planned retreat, government breaks the cycle identified above where community expectations reinforce government behaviors that expend resources on directions that may prove futile in the future. Rather, community expectations are slowly directed toward an understanding that future sea level rise may require an abandonment of the coastal area. Because this expectation is built incrementally into the community through the techniques identified above, the community members are better prepared to make informed decisions about the extent to which they are willing to privately invest in the area. Some might argue that it is a more responsible government that provides a policy setting in which private decisions account for a future probability rather than government conduct that entirely discounts the potential for sea level rise, thus leaving a potential mess for future generations.

Earlier in the section on *staying* at the coast, the categorical techniques of preventing sea level rise and adapting to sea level rise were presented separately but it was also mentioned that these techniques may be used in conjunction with one another to form a more programmatic approach in adapting to sea level rise while remaining along the coast. The examples provided above on *retreating* from the coast were not categorically separated as different strategies, as in the earlier section on staying, but it was mentioned that retreat techniques (such as compulsory and voluntary measures) can be combined to help form a more programmatic approach to retreat as a major policy goal. What has not been explored yet is whether *both* staying and retreat strategies can by utilized as a way of responding to sea level rise.

While it is intuitive that some of these techniques are complementary, it is worth discussing how they might be used in conjunction with one another to create a comprehensive strategy in the coastal zone. The topic of *hybrid management techniques* is the focus of the next topic.

3.3.4 Hybrid Techniques: Combining Staying and Retreating Management Options

Similar to combining armoring and adaptation techniques to provide a wider array of policy options when deciding to stay at the coast, both the options to stay and retreat can be combined into a policy framework providing a combination of choices that are triggered upon the occurrence of certain events. The basic goal surrounding a mix of stay and retreat strategies is to recognize the variety of situations presented in today's coastal areas and the need to provide different sets of options, depending on the circumstances presented. Recall from Chapter 2 that coastal areas can be defined through various criteria, including the extent of human development. Manhattan in New York City was identified as an area that presents a significant commitment of resources within the coastal area based on the existing density and intensity of development. The kinds of planned retreat strategies identified above will have limited impact when applied to an area like Manhattan because of the extent to which it has already been developed. Thus, strategies geared toward remaining in the coastal area might be necessary as an initial policy approach, but they can also be supplemented with retreat strategies that are developed to be implemented both today and in the future should certain events trigger their application. This is but one example where a hybrid approach employing both stay and retreat strategies might be used.

A variety of combinations exist in which both stay and retreat strategies may be combined to formulate a policy framework that plans accordingly for sea level rise. The combination of the two approaches can be accomplished in a programmatic way that accommodates important factors such as current development, future development, and likelihood of impact from future sea level rise at different ranges based on predictions and best evidence. For example, a pre-built area may require armoring techniques in the near term in order to establish a baseline of protection for the coastal population that exists in the region today. This armoring technique may be combined with *future-oriented* planning that looks to provide planned retreat opportunities should sea level rise occur to the point where retreat is the only viable option. In this instance, one is attempting to strike a balance between expectations; protection is being used to reinforce the expectation of *staying* at the shore in the near term, while planning is being done to allow for ease of *retreat* in the future should conditions develop that require this outcome.

The kinds of combinations that may occur between staying and retreating are many, and one way to understand these options in context is to consider three important variables that influence the choices to use any particular combination of techniques. The three variables are as follows: *timing considerations*, *condition of the coastal area*, and *uncertainty* (U.S. Climate Change Science Program 2009, p. 87–104). Each of these variables influences the choice of techniques employed, whether one goal (staying versus retreating) will dominate, and also how the techniques chosen will influence public sentiment and expectations; a well-planned policy scheme that includes a goal of retreat long term may need to consider the kinds of protection techniques employed in the short term from the standpoint of the impact the kind of technique chosen may have on the coastal community's future expectations.* This balancing approach can be achieved through a framework of planning that incorporates the three factors above in a way that clearly establishes a transitioning of priorities over time to best reflect community values while ensuring public safety in the coastal zone. Because these variables are important for planning hybrid approaches to coastal management of sea level rise, they are each explained below in greater detail.

Timing is meant to refer to the consideration of *when* to employ different strategies and/or techniques of sea level rise adaptation approaches. The consideration of timing is really meant to reflect the idea that adaptation policies are dynamic and can change based on factors relevant to time. As mentioned above, an already-developed coastal community may require an immediate policy approach that favors armoring techniques and adaptive strategies providing mitigation support for the community to remain in the coastal area in early stages of sea level rise. Depending on the local conditions, some retreat planning techniques may also be introduced to allow for an ease of retreat in the future on the triggering of certain events. In addition, the early retreat planning techniques (those providing buffer zones between development and the ocean in particular) can support a goal of remaining in the coastal area for a longer period of time during the early to mid stages of sea level rise. In another example, a coastal community may initially engage in retreat techniques early on, particularly methods that provide for buffers between development and the tideline. However, future sea level may come to a point where the buffer zone has been overtaken by the encroaching sea, leaving the community

* For example, a choice to allow armoring techniques to protect short-term coastal interests may include the use of soft-armoring techniques rather than hard-armoring techniques. The preference for soft armoring is made because the technique provides a level of protection today but also limits the expectation that protection will be available tomorrow. As such, the community expectations about a future commitment to armoring may be lessened where soft-armoring techniques are employed rather than hard-armoring techniques. This is particularly true where retreat-oriented policy techniques are employed contemporaneously with the armoring technique; the coastal community is provided with immediate support to stabilize current conditions, but the expectation about future protection is diminished through the combination of policy techniques chosen.

with no buffer between the ocean and development. If the coastal community has not engaged in other retreat techniques as the sea has enveloped the buffer zone, then it is possible the community may choose to switch its strategy from retreat to protection through armoring. Depending on the level of development in the particular coastal area, including the expectations and economic commitments that have been made, it is possible that retreat strategies may be altered to armoring where important assets are put at risk through ocean migration landward.* Yet another slightly different take on timing includes an extensive proactive approach in terms of planning where armoring is provided in the present time but a "deadline" on the armoring is preplanned through far-reaching regulation that includes the adoption of rolling easements to take effect at some designated time in the future (50 to 100 years as example markers for current sea level rise predictions). The effect of this policy choice would be to develop a single instrument that incorporates armoring but also acknowledges the cessation of this kind of commitment within the coastal zone after an extended period of time to accommodate future sea level rise (Titus 1998).† By establishing a date upon which armoring is no longer an option (rather than remaining silent), the community's expectations can more easily reflect this hard deadline where the policy approach changes from protection to retreat.

The *condition of the coastal area* suggests a more holistic planning approach that considers the impact of the technique employed to adapt to sea level rise on *both* the intended coastal area and the surrounding parcels of land. The goal here is to consider the impact of the technique not only on the recipient area, but also on the larger area in general, both in terms of human and non-human responses. For example, the choice to employ a hard-armoring technique to hold back the sea may protect the immediate coastal area, but it also has impacts on surrounding parcels of land that do not include the immediate coastline. For one, wetland interactions may be degraded, particularly where a seawall cuts off water-based interactions between the ocean and the wetland area; estuaries may be cut off from tidal flushing that provides important nutrients to the estuary while also diluting and removing excess waste product into the open ocean. Thus, the choice to hard-armor can

* This example provides some evidence of the importance of managing community (human) expectations when considering policy goals related to sea level rise adaptation. If the community assumes that early buffer zone requirements *are* the whole solution to the sea level rise problem, then the community may engage in ongoing financial commitments in the coastal zone, regardless of the extent of sea level rise observed. The critical time in this example is where sea level rise encroachment into the buffer zone is being observed. The ability and will to reinforce community expectations that include a continual retreat beyond providing the initial buffer zone is necessary; otherwise the community expectations may reasonably change to that of protecting the assets developed, maintained, and extended in the coastal zone during the period between observed sea level rise encroachment into the buffer zone and the eventual submergence of the entire buffer zone under water.

† The kind of *armoring* that would be allowed in this type of policy approach would likely be limited to soft-armoring techniques so that the rolling easement transition would be fully capable of taking effect at the designated point in the future.

have consequences on wetlands that are not immediately adjacent to the coastline but share important connections with coastal processes.

Another example where the condition of the coastline matters in choosing between policy instruments is the use of nourishment techniques; nourishment of sand along a coastal area that has strong coastal dynamics can have different results than nourishment projects along low-energy coastal areas. Providing sand in a discrete area along a beach facing the ocean may have limited impact where the tidal interface contains strong energy dynamics including powerful longshore currents and riptides; the sand provided to bolster the beach area can be quickly transported downshore by the longshore current movement, or otherwise moved offshore through riptide action. Contrast this situation with a low-energy estuarine or bay coastline, where transport mechanisms are much less likely to impact a nourishment project in such areas. Thus, the dynamics of the coastal area are important considerations when contemplating an effective policy approach; nourishment projects along high-energy areas likely need to be expanded in scope or connected with other techniques (such as combined with the use of jetties and groins) in order to maximize the potential use of the project. The same project in a low-energy area may not require additional techniques to accomplish the immediate goal.

Considering local conditions, both terrestrially and seaside, is important in hybridizing several techniques that may utilize both staying and retreat strategies. Consider armoring in relationship to density of development along a coastal area: the extent of armoring can be customized to the level of development so that multiple objectives may be achieved. For example, coastal areas with low development densities might employ both armoring and retreat methods in conjunction with one another. The armoring may be employed immediately in front of a specific coastal property to protect that property. However, the adjacent undeveloped parcels of land on either side of the developed property may remain unarmored so that coastal features may gravitate inland over time. The extent to which this kind of "patchwork" method of intermittent armoring is effective depends on the energy dynamics of the particular coastal area; patchwork armoring may work best in low-energy areas and may be less effective in high-energy areas because a higher rate of erosion in neighboring plots of land may mitigate the effectiveness of the armoring for the developed parcel of land.

The overall trends of population growth in a region can also impact the conditions that are observed in an area and therefore influence the kinds of policy techniques employed because the goals might vary depending on relative demand in the region (U.S. Census 2011). For example, a coastal area that is already developed may have limited capacity for retreat options where the inland areas immediately adjacent to the coastal region are also developed. It may be easier to implement a retreat policy in areas where there is sufficient undeveloped land immediately inland of a coastal area that is most at risk for inundation. In addition, economic considerations may justify armoring in areas where there is a strong preference for living in the region; areas that are experiencing less demand for growth may provide an easier condition for implementing retreat policies because there is less general demand for current and future growth in the region. These are just a few of the variables that apply when considering the condition of the coastal area: intensity of development, characteristics of the coastline, and demand for living in the region. Depending on the mix of factors presented, a variety of techniques that employ both staying and retreat goals can be utilized as adaptation strategies.

Uncertainty was discussed earlier in this chapter in the context of policy development: uncertainty as to the extent of future sea level rise impacts the ability to decipher between policy choices that favor either staying or retreating. The role uncertainty plays in providing hybrid approaches that embrace both staying and retreating is highlighted here through examples where uncertainty can favor both the protection of the shore as well as retreat from the shore. For example, proactive capacity building can always be employed in coastal areas to protect against the kinds of dangers associated with coastal living (flooding, storms, etc.), regardless of the future extent of sea level rise. These techniques can include regulatory setback requirements, rolling easements, and building standards that work together to protect against known risks of coastal living and also provide the capacity to deal with sea level rise in the future should it occur. Developing capacities that make it easier for a coastal community to retreat in the future does not mean the community is prevented from employing techniques to remain under certain conditions. For example, a coastal community that embraces the establishment of buffer zones between the water and development might nevertheless be in the position to enact armoring techniques in the future should sea level rise threaten values within the community and the choice to armor reflects a judgment about the relative costs and benefits of those values in the context of the time in which the decision occurs. In this way, including a capacity to retreat from the area today is not seen as a prohibition against armoring but, rather, such capacity ensures the community greater options in the future depending on the extent of sea level rise actually observed.

3.3.4.1 Zoning Overlay District Example

Before moving on to Section 3.4, which focuses on considerations that might apply among adaptation approaches, an example of the kind of planning instrument that can inculcate the use of hybrid approaches to sea level adaptation is offered. A *zoning overlay district* is a kind of planning instrument that allows a variety of sea level rise adaptation techniques to be employed throughout space and time (LaGro 2008, p. 309–310). In essence, the district provides an additional layer of zoning (land-use regulation) that is "overlaid" on top of existing zoning requirements. Generally, the zoning overlay district is defined by certain features deemed important to coastal management goals, with an emphasis placed on how those features are related to the coastal management goals identified. In the context of sea level rise, a zoning overlay district might be established to help achieve a multifaceted goal of *adapting* to sea level rise. Thus, the planning of the overlay district would need to include a coverage area that is sufficient to meet the multiple objectives involved in sea level rise adaptation. Because planning for sea level rise includes thinking about the future under uncertainty, the zoning district would need to incorporate degrees of uncertainty as part of its planning process; the actual area covered within the district would need to encompass a worst-case scenario of sea level rise, while the particular policy instruments incorporated into the overlay district would need to be flexible enough to account for the uncertainty of future sea level rise. Thus, the policy instruments (tools) contained in the overlay district planning process would likely need to accommodate strategies of both staying and retreating, as these strategies reflect various uses and preferences related to the coastal area that may switch in priority over time based on changing conditions. To better understand how this process would work in context, an example planning process incorporating a zoning overlay district is outlined below.

A zoning overlay district generally begins by engaging in a *planning stage* where the geographic extent of the area to be covered by the district is determined. This is a critical phase because it requires consideration of not only ocean-based impacts on the coastal region today, but also the potential extent of impacts that may occur based on best estimates of sea level rise over a certain period of time (say, 50 to 100 years).* The reason the planning stage is critical is because it establishes the capacity for the overlay district as an instrument to cover the scope of sea level rise impacts in both space and time. For example, if an overlay district is established but limited to an area that is immediately adjacent to the coastline, then circumstances may limit the ability of the overlay district to manage expectations regarding sea level rise. The overlay district will only manage expectations in the area defined, and may even create an expectation that areas outside the boundary of the district are not in danger from sea level rise and thus actually spur activity reinforcing commitments to the area when it may be irrational to make such commitments.† Because the overlay district boundaries will help define expectations, it is generally better to apply the district boundaries broadly to cover all eventual features that may be impacted by future sea level rise. The types of controls that apply to the district can

* These estimates of the extent of sea level rise locally should be based on the best scientific evidence available. Because science is constantly refining its forecast of future events (more data over time reducing the degree of uncertainty), the extent of the overlay district should be updated on a regular basis so that it reflects the best scientific evidence at a given point in time. As such, it is possible that initial boundaries of the overlay district may expand or contract as new evidence sheds light on the potential areas impacted by sea level rise. This kind of policy approach, where the boundaries of an area may change over time to reflect a decrease in uncertainty, is often referred to as *adaptive management.*

† A reasonable perception can be created that the overlay district has incorporated the entire set of risks that can occur from sea level rise. Thus, if one's property sits outside the overlay district, then it is reasonable to assume that investments can be made in that area without consideration of the potential limits to the land that may occur based on future impacts from sea level rise.

be moderated spatially and temporally so that areas within the district but furthest away from the coastline are offered more options for the use of their land in the near term, incorporating "triggering" events where future changes in conditions may prompt greater limitations on lands further inland that are covered by the district.[*]

Once the area of a zoning overlay district has been established, the next step is to determine what techniques may be used within the district to manage expectations of those living in the coastal community. Critical to this planning stage of policy instrument development is to have a full understanding of the various values that are at issue from sea level rise, including the potential to prioritize those values so that policy techniques may be applied in a manner that allows for maintaining as many values as possible (ensuring the highest net value) while still being able to make choices when two or more values may be in conflict (Axelrod 1984; Bales 2001).[†] The choice to provide hard armoring in a coastal region that is subject to sea level rise is one example where multiple values are potentially in conflict with one another. This is mostly true where the area considered for armoring currently has a full set of coastal features (including coastal wetlands) and sufficient upland areas to accommodate the movement of those coastal features inland as the sea level rises. The use of hard-armoring techniques can certainly impact the ability of coastal features to move inland, as discussed earlier in this chapter. Thus, the choice of armoring as a technique to be employed within the overlay district must be considered in light of the conditions presented. The factors that might be considered in this analysis include

- The current extent of development within the coastal region
- The existence of coastal features; the present ecological value of those coastal features, including the provisioning and regulating services currently provided by the features
- The capacity for those features to move landward
- The likely future extent of sea level rise in the area

Considering these factors in relation to one another and also in relation to a priority listing can be useful in helping to determine whether armoring techniques

[*] For example, the furthest inland reaches of the zoning overlay district may provide no *additional* regulation above and beyond existing zoning for the area. However, the fact that the area has been identified as within a *zone of influence* for potential future impacts provides some immediate expectation for landowners in this area that their property may become subject to sea level rise impacts in the future. As more information provides a better understanding as to the extent of sea level rise in the future, the expectations of private landowners can be further managed by incrementally increasing the limitations on the use of land in that zone. This planning technique generally follows the notion of a *rolling overly district,* discussed in greater detail later in this section.

[†] Addressing value-based judgments between policy choices is the purview of the following section where categorical values stemming social, economic, and ecological areas are considered.

will be used, whether any limitations on those armoring techniques (soft armoring only, patch armoring in low-energy areas, etc.) will be utilized, and whether alternative approaches exist that strike a clearer balance between the values of the coastal area.

The techniques chosen to manage a coastal region under a zoning overlay district can be spatially distributed, where stay tactics might be employed in greater frequency immediately along the coastal area for existing development, while an emphasis on planned retreat is utilized to manage pockets of undeveloped coastal and inland areas within the influence of potential sea level rise. The concept here is one where the overlay district provides an equitable distribution of adaptation techniques, acknowledging the commitments made by those who have already developed within the coastal region but establishing an expectation for those who have not developed or future persons who have not yet come to the coastal area. Existing development may allow for techniques providing a level of protection (and thus insurance) over immediate threats to sea level rise, while the district discourages future development or otherwise limits the expectations related to future development by limiting the use of those lands. In this context, some armoring and other protection-focused strategies may be available to existing development but not for new development within the district. Armoring may be allowed along the coastline for already-developed structures that meet specific criteria for armoring.* Rolling easements may be enacted along a coastal area to help create a buffer zone near the coastline. Setback requirements based on erosion rates of "critical zones" nearest the coastline may further limit development. The rolling easement and the setback can reinforce the need to allow for sea level rise to move landward, while limited armoring provides a compromise to investments made by current landowners along the coastline. There are a number of ways in which stay and retreat techniques may be combined to achieve the immediate and long-term goals established in the zoning overlay district.

One of the keys of the overlay district is to ensure that it is capable of evolving with changes in the environment over time. This may be accomplished by providing zones within the overlay itself, where critical areas are defined by proximity to the shoreline, with the greatest restrictions applied in this zone, whereas less critical zones are established further away from the shoreline to the outer extent of the

* The criteria for allowing the armoring might consider the investments made by the current landowner in conjunction with an assessment of the overall impact of the armoring for the specific location on the public resources and other values in light of the multiple goals established for the overlay district. Where the allowance of armoring is shown not to significantly impact the overall policy goals, it may be justified and thus allowed (particularly where a rolling easement may limit the effect of armoring for the individual landowner). In instances where the armoring conflicts with the overarching goals of the overlay district, it may be denied. Critical to the success of such an allowance program is ensuring that the basis for either allowing or denying the application to armor is based on well-established policy goals and criteria that can survive a legal challenge (discussed in greater depth in Chapter 4).

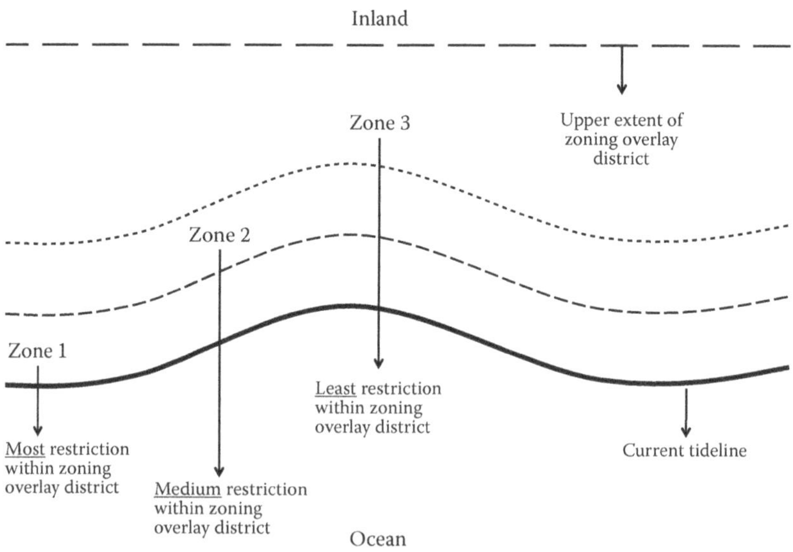

Figure 3.6 Representation of coastal zoning within an overlay district.

overlay district. A representation of this kind of zoning within an overlay district is shown in Figure 3.6.

To make the overlay district relevant in the future, it can have the ability to move landward in conjunction with sea level rise, much like the process that describes a *rolling easement*: as sea level moves inland, the boundaries within the overlay district also move inland. The goal is to ensure the areas of land that might be subject to sea level rise are constantly identified so the limitations expressed in the overlay district can be transferred to new property in conjunction with sea level rise. Recall that the least amount of restrictions would apply to the property at the landward extent of the district. However, the expectations on future uses of land become incorporated into this property *should* sea level rise continue its march landward over time, ensuring that property owners are given sufficient notice and opportunity for their behaviors to begin reflecting the potential of sea level rise.

The purpose of this zoning overlay district example is to provide a summary of how the various techniques employed to deal with sea level rise (via a hybrid approach) may be used in a programmatic sense to control expectations in areas that may be impacted by sea level rise over time. The message of the zoning overlay district is one where planning is a key component to preparing for sea level rise. The kind of planning necessary must include consideration of the spatial boundaries of land that may be impacted by sea level rise over time, and then to moderate that potential for impact using policy instruments in a way that provides *notice* to those where the probability of sea level rise impacts is currently low (the landward extent of the district), while also actively managing the immediate coastal area in a more holistic manner by identifying and prioritizing the many values that are at stake through sea level rise, including choosing policy techniques that balance these values where appropriate and even differentiate between competing values when necessary. The various kinds of adaptation approaches discussed in this section offer a number of techniques that can be applied to achieve a particular outcome based on preferences expressed. Of course, in order to decipher between choices, the values that underlie the preference must be known. Choosing to build a wall against the incoming tide makes sense when thinking primarily about protecting the *present* human development that exists along the coast. However, building that wall may make less sense when the impact of doing so is considered on the *future* human development in the area when projecting future sea level rise. In addition, even the present decision to build the wall may be less appealing when considering other values, such as losing the features of the coastal zone as a result of the wall, including the important habitat that defines many a coastal area. Thus, the question of what kind of technique to use becomes more difficult to answer when that question includes consideration of values among choices, the subject of the next section.

3.4 Considerations among Adaptation Approaches

The question of choosing between potentially competing values is the final topic taken up in this chapter. Often, the overriding issue in policy development does not focus on what *can* be done, but rather what *should* be done under the circumstances (Barrow 2006). When multiple values are identified that may be impacted by a policy decision, a conflict can arise in understanding how to choose among these values (Layzer 2011). One of the reasons for this is that values are dynamic and subject to interpretations from both the individual and group standpoint. Often, values reflect *preferences*, which themselves are based on criteria that can be weighted differently, depending on who is engaged in the valuation process, existing biases that can influence the relative weight given to one criteria over another, and other factors that may yield different outcomes (Binmore 2007). One way to understand how values may differ is to present a representative example so that different preferences can highlight how alternate outcomes can be achieved by emphasizing different preferences on the single criteria under review.

Consider *time* as the single criteria of choice in evaluating multiple options related to sea level rise. The time horizon used to evaluate different policy options (including the particular techniques chosen to employ those options) helps to elucidate how preferences can yield multiple goals and outcomes. If one is focused on today without any consideration of tomorrow, then preferences will tend to reflect policy choices that favor benefits today while discounting costs of those actions tomorrow. In terms of adaptation policy to sea level rise, current observations would tend to dominate the policy tools chosen. Thus, if sea level rise poses no immediate observable threat to a coastal area, then the preference for today would favor a wait-and-see approach; spend little to no resources on sea level rise until it is readily observable and action must be taken. This may be a superior choice from a direct economic standpoint because it ensures a policy that allows a fuller use of the coastal area (less regulation on how the coastal area can be used) until there is a direct and observable reason to alter those present uses. However, the choice to weight today over tomorrow can also have negative outcomes, particularly in the case where sea level rise does occur in the future. Without planning today for the future potential of sea level rise, there is no expectation for this outcome in the coastal population. This means the population can make "bad" investments in the coastal area in the near term that are meant to extend beyond the useful life of the coastal region due to future sea level rise. It can also commit the public entity to limited policy choices in the future. By waiting until sea level rise creates an observable problem on the coast, the available policy choices have likely been limited; it is more likely that the government will choose staying tactics such as armoring to protect against the incoming sea because this is what will be expected from the community. Thus, through armoring techniques, other values (ecological, coastal features) may be placed at risk because there has been no prior planning to include those values in the wait-and-see approach.

The purpose of the example above is to highlight how a preference can impact policy choices; what is rational under a shortened time frame may seem unreasonable for those who prefer to plan for a longer time frame including contingency planning in their analysis. Looking at the above example through a lens where a long-term approach is preferred, the benefits for people living in a coastal community today will likely be balanced against the potential impacts that both action and inaction today can have on the community in the future. While a choice to armor and protect immediate interests may seem a superior choice under a shortened time frame, when the time frame is expanded to include the potential impacts of sea level rise, armoring may have a more limited value. This is particularly true when the expectations of coastal communities are considered as part of the policymaking process; communities are more likely to expect immediate solutions when planning has not incorporated an approach that requires the community to expect sea level rise to change their expectations in how they interact in the coastal area.

Time is one criterion that impacts policy choices, and as shown above there are different preferences that can impact *how* time is applied to sea level rise adaptation approaches. Below, three categories of criteria are discussed—*economic, social,* and *ecological* criteria—all of which can have an impact on planning for sea level rise. Like the example of time above, each criterion is subject to preference choices that can influence how the criterion is being applied to help decide a particular policy direction. These categories are not meant to be exhaustive of the kinds of variables used to make decisions about sea level rise and adaptation strategies along the shore. However, they do represent some of the major themes that are paramount in influencing adaptation choices. As each theme is discussed, recall that preferences help to create a weighted value for each factor. Thus, for example, economic impacts may be subject to different weighting in decision-making depending on how those impacts are being measured. The overall goal in this section is to provide additional insights into how the different policy techniques described above might be implemented, including discussion of how the categorical interests described below apply generally to that discussion and also vary the analysis depending on *how* the category is being interpreted in context.

3.4.1 Economic Considerations

Economic considerations between adaptation approaches are generally seen as the primary means by which policy techniques are chosen (Gerston 2010). Recall the earlier discussion in this chapter on quantifying "net benefits"; the kind of analysis used in determining those net benefits was grounded in economics. Looking at adaptation strategies broadly, the choice to either stay or retreat from the sea has economic consequences—some that are immediate and others that are realized over time. Further, uncertainty impacts a decision on whether or not to incur an immediate cost or apply a wait-and-see approach where costs are incurred only when conditions indicate that the cost is a necessity. If one takes a conservative approach to cost incurrence (only paying for things when it is absolutely necessary), then in an environment of uncertainty, policy choices will tend to be responsive rather than proactive; the choice to take action against a perceived threat will only make sense from a conservative standpoint later in time as the uncertainty of the event is reduced and its likelihood better understood. In the policy world of adaptation to sea level rise, this means actions will tend to move toward a response-oriented strategy incorporating staying techniques as the major policy framework.*

* Uncertainty under a conservative ideology will tend to focus resources on direct economic gains in the near term until the uncertainty is reduced. For developed and developing coastal areas, this likely means reinforcing the existing preferences shown for investing in the continued development and maintenance of the coastal area. As time passes, the continued investments in the developed and developing coastal areas will reinforce policy responses that are protective in nature and geared toward maintaining the investments incurred. This can lead to the situation of protection until "failure," where the potential increases in sea level rise cause conditions that are untenable based on the characteristics of the coastal region. In such cases, a retreat strategy may be inevitable at some point, but the retreat may not be planned (or as well planned) as it may be under a proactive policy approach. Note that the discussion here is on *existing* developed and developing coastal areas where a clear preference for human development and use has been expressed. *Undeveloped* coastal areas are not included in this analysis because, as of yet, there has been no expression of preference for the development of those areas. It may be that certain coastal regions contain sufficient undeveloped land surrounding development such that mitigation of sea level rise may be possible without any planned intervention.

A response-oriented strategy can be contrasted with a proactive strategy that reflects a more liberal economic viewpoint. As suggested earlier in this chapter, proactive strategies choose to incorporate the *potential* future risks into policy management decisions today. As a result, choices may be consciously made today that reduce immediate economic gains (e.g., by limiting the amount and intensity of use in a coastal area), but ensure that the impact of sea level rise (should it occur) is minimized on the coastal community. A proactive policy stance also helps manage human expectations in the coastal area, which can result in less investment where greater investments would otherwise occur. In a way, proactive approaches accept the retardation of potential economic gains in the near term to hedge against the potential harm that may occur over time. The "gamble" in this hedge is that sea level rise does not occur to the extent where it poses significant risk to the coastal area in the future (or technology is enhanced to more easily mitigate the impacts in the future) and the conscious choice to retard growth in the near term is ultimately determined unnecessary. Consequences include less economic output and a lower "well-being" in the near term for current users of the coastal area because preferences have been altered through government intervention in the form of the proactive policy instruments chosen.*

The broad economic question that is the focus of this section is: Which viewpoint is superior for adaptation planning: a conservative viewpoint or a liberal viewpoint? Like the policy instruments available to carry out each viewpoint, there are pros and cons to each approach, and the ultimate decision is often a value-laden enterprise that incorporates some elements of both conservative and liberal viewpoints. While the subjective aspects of the value questions cannot be answered here, some insights may be gained by looking at the different categories of "value" upon which economic decisions generally are made. A *total value* approach is the focus of the remainder of this section, and the term represents an accounting framework where different categories of values can be placed so that a consistent approach to understanding the impact of decisions on these different values may be made. There are two categories of value expressed within a total value approach framework particularly important in sea level rise adaptation approaches: (1) *direct values* and (2) *indirect values* (McGuire 2012, p. 91–100). Each category of value is summarized below and then combined to show how a total value approach works in helping to frame the economic consideration of different adaptation approaches. The overall

* The extent of economic harm to current users of the coastal area can be mitigated by the manner in which the proactive policy approach is implemented. For example, current uses of the coastline can be maximized to the extent possible while still inculcating the expectation of changed behavior patterns under conditions of sea level rise. Well-planned zoning overlay districts that include rolling easement-type instruments are ways to provide an expectation of risk-shifting that does not become operative until sea level rise actually occurs. While this does allow a greater degree of current use of the coastal area, it may not provide the full range of preference-induced behaviors that might otherwise occur under a strategy that reinforces long-term use of the area (such as a publicly financed hard-armoring project).

purpose of the total value approach is to ensure, to the degree practicable, that policy choices are being made based on the best available information. By identifying the major categories of "value" from an economic standpoint, the potential range of impacts of a policy decision may be better understood and therefore the tradeoffs between policy approaches clearly acknowledged. Thus, even where preferences prioritize one category of value over another, there is a greater likelihood that those preferences will internalize the tradeoffs identified in the accounting of all values at stake in the management approach.

Direct value refers to the immediate benefits that are being bargained for in the exchange of goods or services (Millennium Ecosystem Assessment 2003). In the context of coastal zone assets, direct values are those values that are clearly linked to resource investment. For example, coastal land has a direct value based on the purchase price of the land for human use. The "fair market value" of the land is often an approximation of what the land is worth. As previously mentioned, coastal land generally captures a higher price per unit of area than similarly situated land further inland. The willingness to pay higher prices for coastal land is based on a variety of factors, including the preferences of those who are willing to pay higher prices to live in close proximity to large bodies of water (Keohane and Olmstead 2007). The kinds of values associated with living close to the water, when they are part of the market price paid for coastal proximity, may be said to represent the "direct" values associated with the land. It is not only land that is valuable along the coast; views are also important attributes of coastal areas, and the "value" of the view can be reflected directly through a willingness to pay for that view. A piece of land that contains unobstructed views of the ocean often commands a premium price over adjacent land where the view of the ocean is obstructed by development between the land and the water. Some coastal property owners have been known to purchase adjacent developed property in order to simply tear down the structure impeding the original lot's view, thus internalizing the additional value of the view into the original property through whatever price was paid for the adjacent developed land and demolition of the structure on that land.

The concept of direct valuation contains a few assumptions worth noting. First, there is an assumption that the price being paid reflects complete information; both the buyer and seller have equal access to all information necessary in order to accurately price the transaction, and the price paid for the transaction, on average, internalizes all this information. However, this assumption is often incorrect (Bradfield 2007, p. 29–31; McGuire 2012, p. 63–108). There is often incomplete information available to both buyer and seller when settling on a final price in a transaction. For example, a buyer may be willing to purchase a coastal property for a price that assumes the property will be capable of development and (essentially) uninterrupted use for a period of time. If sea level rise and associated regulation of coastal land in the area prevents this assumption from coming true, then the price paid by the buyer is based on incorrect information.* A second important assumption about direct valuation methods is that because this method includes market-based transactions, the actual price paid is reflective of all the values that are to be found in the underlying resource that is the subject of the transaction. So, in the coastal land purchase and sale example above, the ultimate price paid and accepted for the coastal land is somehow reflective of all the values that are contained in that coastal land. However, this assumption often does not hold true when identifying values that are not included in the transaction. For example, the coastal land may provide services beyond that as substrate for human development; the land may be important for the regulating services it currently provides along the coastal area. The kinds of values that may be attributed to the land outside of direct valuation techniques must also be considered in determining its overall value and worth.

Indirect value refers to values that may be found outside of traditional market-based direct values (Kopp and Smith 1993; McGuire 2012, p. 95). Indirect values are meant to represent value that is often not considered in the market-based transaction. Using the example above, the indirect values related to coastal property can include storm protection, groundwater protection, ecologically critical habitat, water filtration, and many other services the property may provide in helping to regulate background environmental conditions. The importance of indirect values is that they can help add context to a policy debate on how best to approach coastal management options. For example, a coastal area that has significant nearshore coastal wetlands providing regulation of water quality and nursery grounds for important commercial and recreational fish species places a policy choice about armoring into a different context than if the choice is *only* based on protecting human habitation. While the cost of the armoring project may be justified when compared to the direct values of the coastal development in the area, the loss of coastal features and wetland functions from the armoring project can change the overall analysis and ultimate conclusion on net benefits from the project. By considering indirect values in the policy analysis, there is a greater chance of capturing additional values outside of those that exist solely in a market-based transaction.

Collectively, direct and indirect valuation methods can be used in sea level rise adaptation planning as a way of measuring the provisioning and regulating services of the coastal area under consideration. Once understood, these categories of value

* Government can always choose to provide a level of assurance to these kinds of transactions by engaging in actions that help bolster certainty about future events. A publicly funded project to hard armor the area against the shoreline can help increase the likelihood that the area will be protected from sea level rise for a sufficient period of time. On the other hand, a government-initiated policy that proactively states no armoring of the shore will be allowed in the future can help both buyer and seller in adjusting the relative value of the property in light of the policy to allow coastal features to move inland. In either case, government action can help provide a level of certainty in private transactions and thus help determine the direct value of the land under consideration.

can be compared to different policy techniques. For example, the impact of large-scale armoring projects can be compared to nourishment projects, which can also be compared to buffer zones, setback requirements, as well as hybrid mixes of techniques. The different outcomes presented by applying the several techniques can be compared to the categories of value identified so the impacts of varying techniques on the categories of values can be realized. Once the relative benefits and costs of different techniques have been identified and compared, a larger-scale goal, such as "net benefit," can then be applied by comparing the relative costs and benefits. The key is in establishing the full range of values that are present and potentially impacted in the coastal area. Focusing entirely on direct value may miss important benefits that exist in the coastal area through indirect values and thus result in an inaccurate identification of values. Alternatively, by engaging in a total value analysis, a coastal area may be found to lack a number of indirect values. In such cases, techniques that favor direct values over indirect values may make greater sense than in coastal areas that have a high degree of indirect value identified.

Considering *economic* benefits and costs between adaptation approaches means considering, fundamentally, how *all* benefits of a coastal area are weighed against *all* costs of a certain action (or inaction). To judge the relative costs against benefits, the total values associated with a coastal area must be identified and, to the extent practicable, quantified. Often, indirect values cannot be easily quantified because they are not regularly included in market-based transactions. Thus, there is no direct measurement of these kinds of values. However, there are ways of measuring indirect values in a coastal area. For example, the value of a buffer zone for coastal protection purposes can be estimated by looking at the potential impacts of coastal storms in the area, the value of properties, and costs of rebuilding, and then determining what kind of development density might exist in a coastal area if an undeveloped buffer zone of land between development and the sea did not exist. Making some reasonable assumptions about how this buffer zone would act to mitigate coastal storm damage, a calculation could be made as to the relative value of the area left undeveloped as a buffer against coastal storm protection (including taking into consideration lost revenues from nondevelopment and other factors that would limit the positive value of the area such as coastal protection). Other indirect values are not so easy to quantify but there are tools available to help determine those values. For example, certain coastal areas may serve as important habitat for marine species that are important sources of tourism for the area. A *contingent valuation* method can be employed to help determine the existence value of that species if people are drawn to the area to view the species in its natural habitat (McGuire 2012, p. 100–101).*

* The kinds of value used to help identify the existence value of the species may include the costs paid by tourists to fly and stay near the coastal area, as well as the costs paid for activities in the area to see the particular species.

While imperfect, indirect valuation methods can be used in combination with direct valuations to help provide a more complete picture of *value* associated with the coastal area. Once this value is identified, the impact of different policy options becomes more coherent. Choosing to stay in the coastal area versus using a retreat or hybrid approach is now grounded by the impacts these different techniques will have on the coastal region. Other factors, including intergenerational dynamics, may also play into the decision-making process; the decision about how to manage the coastal area may extend not only to a full consideration of value, but also to a consideration of the impact of these decisions on those who will inherit the decisions being made today. Economics taken broadly as a framework for analyzing among adaptation approaches can be useful when employed holistically to a full accounting of values. Even when certain values are discounted in the decision-making process, a full accounting ensures that the discounting being done is a conscious exercise where tradeoffs are being considered between approaches; this is the major benefit of engaging in a total value economic analysis—transparency in decision making.

3.4.2 Social Considerations

Recall earlier in this chapter the representation offered in Figure 3.1 of a schematic suggesting the environment establishes the outer boundaries from which human interactions occur. Within the environmental boundaries of the Earth system are social boundaries. Within the limits of society are economic activities. Thus, the representation in Figure 3.1 suggests that our economic activities are constrained by our social institutions, and also that both our economic activities and social institutions are constrained by the environment. The preceding discussion about economics suggests the need to understand how policy proposals impact environmental boundaries by accounting for those values (indirect values through regulating services) when considering techniques to employ in achieving a policy goal of sea level rise adaptation. In this section, the role of social institutions is summarized to help provide a sense of how societal expectations can influence policy directions. Some of these influences have already been expressed earlier in this text. The purpose here is to define these influences more concretely by relating them to considerations between adaptation approaches.

Social institutions are primarily related to the choices made among adaptation approaches by understanding the dynamics that exist between the social institution and members of society (Bales 2001). Here, the term *social institution* is primarily meant to refer to government in its representative forms: legislative, executive, and judicial branches (Rosenbloom 2008). Government has a role to provide *benefits* for its citizens. In the context of sea level rise adaptation, the question of exactly what constitutes a "benefit" is often difficult to discern. Relating the question of what is a benefit to the economic discussion immediately preceding this section, it may be suggested that benefits to society are largely measured in instances where government assesses the relative merits and costs of a proposed activity to determine the action that will achieve the greatest "net benefit" for society. In a world where everything remains constant, the notion of "net benefit" may be easier to determine than in a world where things change because, in a static world, what is deemed a benefit today will generally remain a benefit tomorrow. However, the dynamic nature of our environment—particularly the dynamic nature of the coastal zone—makes the question of what a "benefit" is more difficult to answer. A government action to protect the shore by hard armoring may provide immediate benefits to those currently living along the coastline most at risk due to sea level rise. However, the choice to armor may have more costs than benefits to those living away from the coastline where the choice to armor results in an increased risk of loss to those coastal areas (via catastrophic event), and the cost of that loss is distributed among the larger population of society.* The choice to armor may also have greater costs to those living along the immediate coast in the future should the armoring provide an incentive for those members to remain along the coastline and continue investing in their choice to live along the coast. If future increases in sea level force the evacuation of the area because of a failure of the armoring, the choice to armor may achieve a net loss for current coastal dwellers. Finally, future members of society may be placed at a disadvantage if the armoring proves to remove important environmental functions of the coastal area that could have existed if the armoring was not allowed. What is and is not a "benefit" is variable and depends on factors that include members of society, present and future, being considered in determining the benefit.

In managing coastal areas, it is critical to consider the impact that different policy choices and directions have on different members of society, both in the present and the future. The clearest framework that may allow for the balancing of various interests is through managing societal expectations. The idea of thinking about the expectations of society has been discussed earlier in this chapter, and the goal now is to extend that idea to the consideration of adaptation techniques as they impact social expectations. Social expectations may be linked to behavioral characteristics, at least on a large scale, such that understanding how policy options impact behavioral choices is one way to better comprehend what may end up being a "superior" policy direction, not necessarily because of what it achieves immediately in terms of sea level rise adaptation, but more so because of how it alters social

* Consider the example of a catastrophic loss where insurance premiums are increased for all current members of society in order to help offset the losses incurred by those living along the immediate coast. Imagine a scenario where the costs are so substantial that they also impact future members of our society (so-called *legacy costs*).

behavior patterns in a way that efficiently transfers the risk of future sea level rise to the individual members of society who may be impacted by that risk.*

Consider the following example of how government actions (or inactions) help provide the internalization of risk by individual members living within a coastal area and also ensure the consideration of indirect values (or at least the protection of those values) in the process. Strategies that focus on a planned retreat as part of a proactive planning approach are the best kinds of techniques that balance current uses within the coastal zone while allowing the internalization of risk by private members of the coastal community and thus form the basis of the example provided here. Consider a zoning overlay district as described in the hybrid approach above that focuses coastal planning on three major goals:

1. Preservation of coastal features
2. Maximization of current use of coastal features
3. Internalization of risks by private members of the coastal community

Collectively, the three major goals here are meant to provide a basis upon which coastal planning seeks to balance the uncertainty of future sea level rise with the best current scientific evidence available regarding the likelihood of sea level rise in the area. Consideration is also given to the variety of assets identified in the coastal features so that an indirect value can be ascertained as to the relative current importance of the coastal features present.† An assumption is also made that suggests current coastal features are of value regardless of information available on the specific values represented in the particular coastal setting. Thus, in the absence of specific information, it is assumed that coastal features are important and valuable both today and in the future. Finally, an overriding assumption is one of *precaution* that overlays this entire approach. Precaution means that in the absence of convincing evidence to the contrary, any approach taken should seek to maintain the current status of background conditions because those conditions reflect an environment that has proven conducive for human survival and overall well-being (Crawford-Brown 1999).

* The uncertainty of future sea level rise poses a degree of unknown, but often substantial, risk for coastal areas. The government's role is, by many accounts, to help manage this risk. The suggestion here is that managing risk under uncertainty is to inculcate a sense of this risk in the individual members who choose to live and invest in coastal areas. By providing a setting in which the potential risks are internalized, there is a greater chance that individual decisions will better reflect that risk in market transactions (direct value choices).

† For example, if the particular coastal area contains important habitat for commercial fish species, then that value should be reflected in the accounting of indirect values. Similarly, if the coastal features contain dunes that help protect against storms or wetlands that provide habitat and help to filter water, then those additional features are accounted for in the indirect values of the particular coastal area.

With these basic goals and criteria in place, it is now time to play out the zoning overlay district using different techniques that might further the societal goals by developing an approach that manages human expectations in the coastal zone. First, as noted previously, the overlay district creates a geospatial distinction to the area it occupies; the fact that my property is located in this overlay district begins to alter my expectations. Recall that a zoning overlay district is a special kind of zoning that is added to existing background zoning; the district identifies a particular relationship between the property and the reason behind the overlay district itself. An overlay district indicating land is located within a zone that can be impacted by sea level rise presents an impression with private landowners that identifies their land as existing within a particular zone of influence that may not be readily known or understood by those landowners. Regardless of what occurs in the future, the fact that the zoning overlay district has identified the area as potentially at-risk due to sea level rise creates an immediate awareness of this potential. As such, the zoning overlay district helps to internalize the potential risks of sea level rise in private landowners at the earliest possible time and hopefully well before sea level rise actually impacts the area.

With the overlay district established,* the next step in the process is to choose among adaptation policy techniques that will help reinforce the internalization of risk by private coastal property owners, and also support the other goals of preserving coastal features while maximizing the current use of the coastline. Focusing on the internalization of risks, a variety of planned retreat techniques identified earlier in this chapter may be considered as tools to help accomplish this goal. Buffer zones, building setback requirements, and rolling easements are all techniques that help to reinforce private-party expectations toward the internalization of risks associated with sea level rise. Buffer zones reinforce the expectation that areas closest to the coastline will be reserved for future sea level rise; if the buffer zone essentially "rolls" with the incoming tide (via a rolling easement strategy), then incremental increases in sea level will further reinforce human expectations by moving activities further inland; as the ocean moves inland, the limitations on human activities also move inland. Building setback requirements establish sea level rise as a consideration at the planning stages of private

* There is a variety of factors that can influence the ultimate determination of precisely where to site the zoning overlay district. These factors should be based on the best evidence of sea level rise impacts based on the morphology of the coastal area under consideration. For example, areas deemed highly likely to be impacted by the lowest estimates of sea level rise based on their relative position to existing coastline and sea level may be listed as "high likelihood" impact areas, while inland areas that might only be impacted through worst-case scenario estimates of sea level rise might be listed as "low likelihood" impact areas. Although the factors for defining overlay district boundaries should be based on objective evidence, there is certainly the potential for politics to play a role in the geographic placement of the district due to the potential impacts a designation might have on risk shifting (insurance rates) and valuation of the property.

development by forcing a consideration of erosion rates and how this can impact the future location of the proposed development in relation to the ocean over time. In a sense, the setback requirement, along with the property established in an overlay district, forces private decisions to consider future events that may be *beyond* a time frame usually considered if the zoning district or setback requirement did not exist.

The techniques mentioned above (buffer zones, setback requirements, rolling easements) also help meet the goal of preserving coastal features. All the techniques, individually or in combination, help ensure that the land closest to the shoreline is left in an undeveloped state. Also, the techniques tend to rely on a lack of hard armoring in order to achieve their intended purposes. Thus, there is a greater likelihood that coastal features will have the capacity to migrate inland as sea level rise occurs. In addition, rolling easements in conjunction with mobility building standards can reinforce the expectation that development may have to give way to the incoming tide, depending on the extent of sea level rise. This ensures that coastal features will be minimally obstructed from moving landward over time. Not only do these techniques provide for the preservation of present and future coastal features, but many of them are *contingent* in their planning, meaning they only become operative should sea level rise occur to the point where their more restrictive features are triggered. This ensures a strategy that is both precautionary but minimally invasive toward the current uses of coastal features.

The final goal, maximization of current coastal feature use, can also be achieved through sub-planning that occurs within a zoning overlay district. At the outset, it must be acknowledged that the greatest level of freedom for the use of coastal features will generally be created under a minimal to zero regulatory environment; all things being equal, greater regulation tends to provide greater limitations on the ways in which land may be put to use. Acknowledging this fact, the question really being entertained here is: Under the assumption that regulation *will exist* in the coastal area in one form or another, what is the least restrictive form of regulation that can be achieved that meets the goals outlined above (preservation of coastal features, maximization of current coastal feature use, and internalization of risks) in as equitable a manner as possible? To begin answering this question, managers can look to variants that exist in techniques aimed at a planned retreat while also entertaining some of the techniques (or aspects of techniques) that exist under armoring and adaptation strategies.

Rolling easements provide a longer-term planning technique that is ideal for allowing maximum reasonable use of coastal areas in the present while acknowledging that the conditional event of future sea level rise will change the kinds of uses allowed. In theory, a rolling easement can allow all the primary functions desired of a coastal property in the near term (development, habitation, use) while acknowledging these activities may be of limited duration in the event that future sea level rise touches the property in question. In this way, the *concept* of the rolling easement allows private decision making to maximize current benefits within the coastal zone. If the rolling easement technique is combined with a buffer zone prohibiting development within a set distance between the shoreline and upland, then this additional technique can place further limitations on current uses of some land beyond the rolling easement itself.* The extension of limitations through the buffer zone places additional restrictions on the present use of some land, but does so while managing human expectations in the coastal zone to a greater degree than the rolling easement itself. Yet, both methods are triggered by *actual* sea level rise and thus minimize the uncertainty of sea level rise in the manner in which human activities are controlled.

Certain other sea level rise adaptation techniques may be incorporated into the zoning overlay district to allow the maximization of use in the present while also holding true to the other goals outlined in the zoning overlay district. Nourishment projects can provide stability and a degree of short-term certainty that reinforces current uses until sea level rise occurs to a point where the nourishment is no longer beneficial. Building standards can be employed that allow for structures to be easily relocated outside a zone of danger should sea level rise occur to the point that movement of the structure is necessary. Some degree of jetties, bulkheads, and breakwaters may be employed just offshore to help sustain coastal landmass in critical areas in the absence of significant sea level rise. Also, in certain conditions, soft armoring and other checkerboard armoring techniques may be employed in site-specific areas where conditions will show that armoring provides substantial benefit to the immediate landowner(s) while exacting little to no damage to the indirect values and attributes of the coastal zone itself. All these particular techniques may be employed under a comprehensive management scheme unified under a zoning overlay district.

The above example of a zoning overlay district as a means of managing human expectations as a proxy for social considerations shows how different policy techniques can be combined to allow for both the use and protection of coastal assets under a condition of sea level rise. The zoning overlay district itself helps define overall human expectations in the coastal zone, while the particular techniques employed under the overlay district (including the degree to which these techniques are employed) help achieve the goals of preservation, use, and cost internalization. The unique situations encountered in a particular coastal area will influence the *manner* in which these different techniques are employed to achieve equitable outcomes based on the conditions presented. Areas that have a long history of coastal build-out may require additional protections of human investment to support this legacy, while areas with less of a history can more easily ingest a management paradigm that reduces the long-term expectations of human interest in the coastal area should sea level rise interfere with those expectations.

* The rolling easement would reduce the use of the land when water touches the land. The buffer zone precedes this event and thus may place additional limitations on land before the conditions that trigger the rolling easement restrictions occur.

3.4.3 Ecological Considerations

Ecological considerations among adaptation approaches focus on the importance of indirect values that exist under the "economic" lens of valuation (McGuire 2012, p. 95–98). As noted in Chapter 2, the Earth can be seen as a "system" that operates as a set of processes where energy flows between system components, resulting in an equilibrium, or background state, where predictions about future states are based in large part on what has been observed in the past; the concept of seasons (summer, fall, winter, spring) in temperate regions of the Earth is based in large part on an assumption that the "Earth system" is at a state of equilibrium. Beyond a change of seasons, indicators that the Earth is in "balance" with background conditions that support human well-being include such ecological factors as biodiversity (the amount of species diversity among living things), which itself is dependent on habitat. Relating this presumption to the coastal zone, ecological considerations presume that coastal features exist in an area for a reason, and further presume that the relative "health" of that coastal area is dependent on the preservation of those coastal features to ensure that biodiversity and other ecological processes can be maintained. Considering ecological factors among adaptation approaches, then, is really about ensuring that the techniques applied to coastal adaptation management consider and, to the extent practicable, preserve the ecological features of a coastline.

Because ecological considerations seek to preserve coastal features for the reasons mentioned above, the current state of a coastline is an important starting point in deciphering the potential impact(s) of adaptation policy techniques on coastal features. This is where the discussion on coastal morphology (Chapter 2) becomes important when applying the characteristics of a coastline to a policy technique. For example, a coastline dominated by high rocky bluffs may have fewer ecological values at stake from hard armoring than a low-lying coastline serving as the emptying point for a terrestrial river with a large estuary and associated wetlands that provide habitat for a variety of species.* Armoring the low-lying coastline would retard a number of coastal processes between the ocean, estuary, and river that could lead to physical changes, including the loss of estuarine habitat, thus impacting biodiversity in the area over time. This is but one example where the techniques employed should consider the ecological impacts on the coastal area. And where the precise ecological assets of a coastal area are unknown, the policy framework (as noted earlier) may adopt a precautionary approach that presumes there are ecological values worth preserving.

Considering how adaptation techniques impact economic, social, and ecological categories within the coastal zone helps add some depth of analysis when prioritizing and choosing between policy options. None of the examples provided

* There is really no good reason to hard armor a high rocky bluff because the physical characteristics of the area create a natural "hard seawall." However, the point here is to provide a stark example of how the same technique applied to different areas can yield different results.

here are meant to create an absolute, set-in-stone priority listing that can be simply applied based on the circumstances presented. The reality is that any policy process is much more dynamic and subject to a variety of small data that interact to create overall preferences. However, the orienting processes offered in this section provide some additional categorical considerations that can aid decision makers in creating their framework for determining an overall approach in adapting to sea level rise, and then choosing techniques that provide the best chances of achieving that overall approach in the most equitable manner possible.

3.5 Conclusion

This chapter was fundamentally about the different policy options that are available in making choices about adaptation approaches to sea level rise. The actual options are relatively clear at the largest scale of analysis: a choice can be made to *stay* at the shore, *retreat* from the shore, or a choice can involve a *hybrid* mix of both staying and retreating strategies. What considerations go into making this large-scale choice is not so simple, however; there is a variety of factors that play into these decisions. Helping to complicate matters is the large degree of uncertainty that exists with future sea level rise; the greater the uncertainty, the less likely commitments should be made today that spend limited resources. When uncertainty is reduced, a clearer calculation of risk can be made in order to justify the commitment of resources. In summary, policy options should reflect a framework that seeks to obtain the greatest *net benefit* for coastal dwellers and others affected by coastal decision making.

Looking at the question of policy approaches through a lens of net benefits provides some direction; generally, policies that tend to trigger the bulk of financial commitment when sea level rise is actually occurring (when uncertainty is reduced) are preferred over policies that commit resources under greater levels of uncertainty. In addition, policy preferences can be further refined by looking at important categorical considerations between adaptation approaches—particularly economic, social, and ecological considerations. When net benefit analysis is combined with certain presumptions about economic values, managing social expectations, and ecological integrity, then certain policy techniques tend to be favored over others. For example, retreat strategies that imbue a sense of limited use over coastal property *when the condition of sea level rise occurs* allows for a good deal of present use while ensuring that the bulk of resources is not committed until sea level rise is actually happening; zoning overlay districts employing buffer zones, setbacks, and rolling easements are good examples of such policy techniques. In addition, retreat strategies also help preserve coastal features that may be critical in ensuring important indirect economic values through preservation of ecological integrity. These strategies also help to manage social expectations of coastal landowners by allowing them to internalize the risks of sea level rise in their personal planning at the earliest possible time, creating a proactive environment for coastal resource planning rather than a more reactive environment that may ensue without the application of such techniques.

There may be instances where coastal policy approaches are best advanced through a mix of approaches as defined in this chapter. Hybrid management options allow for the use of a variety of techniques that can be applied based on the circumstances presented. The key lies in understanding that these techniques are "tools" that are available to achieve an overall goal, that is, the equitable use of coastal resources both today and tomorrow. The guidelines established in this chapter are meant to provide a planning framework that highlights considerations regarding the process of decision making that should be engaged in during all planning phases, while also offering a variety of adaptation techniques that can be employed in certain combinations to carry out the final management plan. Whatever plan is ultimately created to manage coastal resources, there is a major threshold that each policy planning phase must go through, and this threshold is a consideration of the legal frameworks that impact coastal zone management. Chapter 4 focuses on these legal considerations and how they impact adaptation management planning.

References

Axelrod, R. 1984. *The Evolution of Cooperation*. New York: Perseus Books.

Bales, R. 2001. *Social Interaction Systems: Theory and Measurement.* New Brunswick, NJ: Transaction Publishers.

Barrow, C. 2006. *Environmental Management for Sustainable Development, 2nd ed.* New York: Routledge.

Basco, D. 2003. Shore Protection Projects. *Coastal Engineering Manual* (Manual 1110-2-1100). U.S. Army Corps of Engineers, Washington, D.C.

Binmore, K. 2007. *Game Theory: A Very Short Introduction.* New York: Oxford University Press.

Bowersox, J. 2002. Sustainability and environmental justice: A necessary connection. In *The Moral Austerity of Environmental Decision Making: Sustainability, Democracy, and Normative Argument in Policy and Law*. Durham, NC: Duke University Press.

Bradfield, J. 2007. *Introduction to the Economics of Financial Markets.* New York: Oxford University Press.

Costanza, R., R. d'Arge, R. de Groot, et al. 1997. The value of the world's ecosystems services and natural capital. *Nature*, 387: 253.

Crawford-Brown, D. 1999. *Risk-Based Environmental Decisions: Culture and Methods.* Norwell, CT: Kluwer Academic Publishers.

Dean, R., and R. Dalrymple. 2002. *Coastal Processes with Engineering Applications.* New York: Cambridge University Press.

Field, B., and M. Field. 2006. *Environmental Economics: An Introduction, 4th ed.* New York: McGraw-Hill.

Gerston, L. 2010. *Public Policy Making: Process and Principles*. Armonk, NY: M.E. Sharpe, Inc.

Hayes, T., D. Spafford, and J. Boone. 2007. Actuarial Rate Review: In Support of the May 1, 2007 Rate and Rule Changes. Federal Emergency Management Agency, Washington, D.C.

Hayes, T., and D. Spafford. 2008. Actuarial Rate Review: In Support of the May 1, 2008 Rate and Rule Changes. Federal Emergency Management Agency, Washington, D.C.

Intergovernmental Panel on Climate Change. 2007. IPCC Fourth Assessment Report: Climate Change 2007 (AR4). http://www.ipcc.ch/publications_and_data/publications_and_data_reports.shtml (accessed August 10, 2012).

Kennedy, L. 1992. *Planning a City upon a Hill: Boston since 1630.* Boston, MA: University of Massachusetts Press.

Keohane, N., and S. Olmstead. 2007. *Markets and the Environment.* Washington, D.C.: Island Press.

Kim, H., and C. Karp. 2012. When retreat is the better part of valor: A legal analysis of strategies to motivate retreat from the shore. *Sea Grant Law & Policy Journal*, 5(1): 169.

Kleppel, G., M. DeVoe, and M. Rawson. 2006. *Changing Land Use Patterns in the Coastal Zone: Managing Environmental Quality in Rapidly Developing Regions.* New York: Springer.

Kopp, R., and V. Smith. 1993. Understanding damages to natural assets. In *Valuing Natural Assets: The Economics of Natural Resource Damage Assessment*, edited by R. Kopp and V. Smith. Washington, D.C.: Resources for the Future.

LaGro, J. 2008. *Site Analysis: A Contextual Approach to Sustainable Land Use Planning and Site Design.* Hoboken, NJ: John Wiley & Sons.

Layzer, J. 2011. *The Environmental Case: Translating Values into Policy.* Washington, D.C.: CQ Press.

Lewis, D. 2012. The relocation of development from coastal hazards through publicly funded acquisition programs: Examples and lessons from the Gulf Coast. *Sea Grant Law & Policy Journal*, 5(1): 98.

McGuire, C. 2012. *Environmental Decision-Making in Context: A Toolbox.* Boca Raton, FL: CRC Press.

McGuire, C., and J. Hill. 2012. Climate adaptation and the Fifth Amendment of the U.S. Constitution: A regulatory takings analysis of adaptation strategies in coastal development with application to Connecticut's coastal management regime. *Sea Grant Law & Policy Journal*, 5(1): 140.

Millennium Ecosystem Assessment. 2003. Ecosystems and Human Well-being: A Framework for Assessment. http://www.maweb.org/en/Framework.aspx (accessed August 11, 2012).

National Research Council. 1987. *Responding to Changes in Sea Level: Engineering Implications.* Washington, D.C.: National Academies Press.

National Research Council. 2007. *Mitigating Shore Erosion Along Sheltered Coasts.* Washington, D.C.: National Academies Press.

Nordhaus, W. 2001. Global warming economics. *Science*, 294: 1283.

Patterson, M., and B. Glavovic (Eds.). 2008. *Ecological Economics of the Oceans and Coasts.* Cheltenham: Edward Elgar Publishing Limited.

Randall, M. 2003. Coastal development run amuck: A policy of retreat may be the only hope. *Journal of Environmental Law and Litigation*, 18(1): 145.

Rosenbloom, D. 2008. *Public Administration: Understanding Management, Politics, and Law in the Public Sector, 7th ed.* New York: McGraw-Hill.

Samuelson, P., and W. Nordhaus. 1989. *Economics, 13th ed.* New York: McGraw-Hill.

Tietenberg, T., and L. Lewis. 2010. *Environmental Economics and Policy, 6th ed.* New York: Pearson.

Titus, J. 1991. Greenhouse effect and coastal wetland policy: How Americans could abandon an area the size of Massachusetts at minimum cost. *Environmental Management*, 15(1): 39.

Titus, J. 1998. Rising seas, coastal erosion, and the takings clause: How to save wetlands and beaches without hurting property owners. *Maryland Law Review*, 57: 1279.

U.S. Census. 2011. NOAA's State of the Coast. http://stateofthecoast.noaa.gov/population/welcome.html (accessed August 10, 2012).

U.S. Climate Change Science Program. 2009. Synthesis and Assessment Product 4.1 — Coastal Sensitivity to Sea-Level Rise: A Focus on the Mid-Atlantic Region. http://www.climatescience.gov/Library/sap/sap4-1/final-report/sap4-1-final-report-all.pdf (accessed August 10, 2012).

Wilson, M., and S. Liu. 2008. Non-market value of ecosystem services provided by coastal and nearshore marine systems. In *Ecological Economics of the Oceans and* Coasts, edited by Murray Patterson and Bruce Glavovic. Cheltenham: Edward Elgar Publishing Limited.

Chapter 4

Legal Considerations

4.1 Introduction

In the United States, the law provides a framework that authorizes certain government action while simultaneously placing limitations on other government actions. This mix of authority and non-authority provides an important conceptual framework from which policy options can be viewed. Conceptually, think of the law as a *box* as represented in Figure 4.1.

The area inside the box in Figure 4.1 represents the authority granted to government under the law. The area outside the box represents non-authority, or a limit to government authority. All activities that are allowable in developing policy directions are limited to the area inside the box; activities that bring government conduct outside the box are unlawful and not allowed. In this sense, the law creates the outer boundaries that limit the kinds of policies that can be considered, created, and implemented.

The purpose of this chapter is to explore *both* the allowances and limitations created in legal frameworks, and how these limitations impact policy directions related to sea level rise adaptation strategies. Often, the easiest policy solution to a problem has legal limits (and political ramifications as well) that prevent a policy from being implemented. This is true even when those legal limitations make it difficult for government to engage in actions that are meant to protect society as a whole. For example, if sea level rise is threatening a coastal region, then we might expect government to enact a regulation that prohibits all human habitation and use of that coastal area. Government seemingly has the legal authority to prevent activities in areas where there is a threat to human safety and welfare, and any regulation seeking to prevent a reasonable harm is certainly within the legal power of government. This is all true, but while there is a legal basis for the

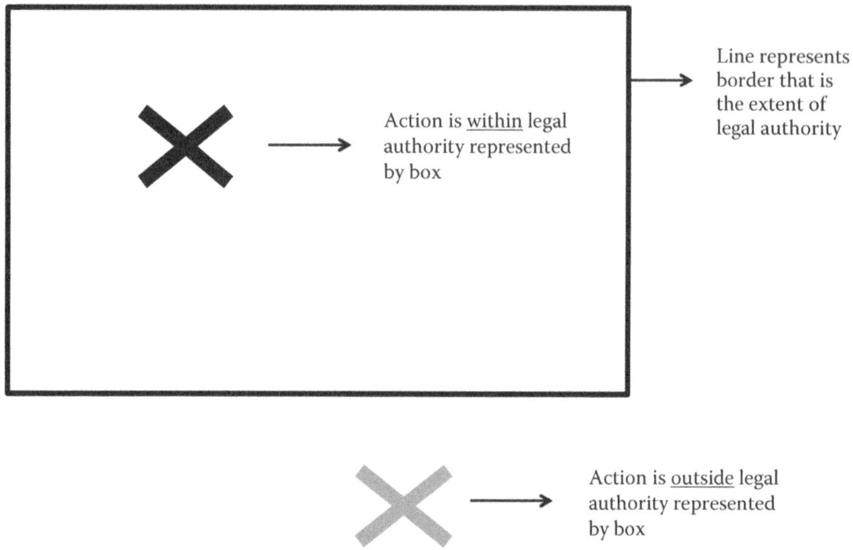

Figure 4.1 Representation of legal frameworks as boxes.

government action, there are also legal *limitations* on how far government can go in instituting such a broad limitation on land use. Consider a government ban that includes land that is privately owned, some of which is developed. The effect of the ban will prevent private landowners from living in their coastal homes, while also preventing undeveloped land from the possibility of development, a property right that might have existed within the title to the land prior to the regulation. While government may have the power to enact the ban based on public welfare (i.e., the *purpose* of the ban is well within government's police power), government may be limited in enacting the ban without first *compensating* affected private landowners the fair market value for the loss of their private property rights. This means government would have to pay for the value of the developed and "developable" land that is subject to the ban. In essence, the government action in banning people from living within the coastal zone is within the legal "box" as represented in Figure 4.1, but the failure to *pay* compensation commensurate with the ban takes the government action outside the box and thus makes the act unlawful without compensation.

The issue of compensation becomes important because it speaks to the initial presumption about the *simplicity* of the policy; the policy prescription of banning human use of the coastal area achieves the goal of protecting the coastal zone, but it does not achieve the goal in the most *efficient* manner when financial costs are considered. The law creates a framework that alters the policy prescription itself because it requires compensation from the government in order to implement the policy initiative. By requiring the government to pay for the individual losses created by the regulation, the costs of the policy proposal are shifted from private parties to a public entity through the legal limitation. Therefore, the mathematics of the policy prescription is altered; indeed, a variety of other policy proposals might exist that are more cost-effective in achieving the goal of protecting the coastal community while avoiding some of the legal issues described above.

In many ways, this chapter is about understanding how to implement adaptation policies in light of limitations on government actions established through legal frameworks. Conceptually, the goal in this chapter is to understand what laws comprise the "box" in sea level rise adaptation management and, through this understanding, determine what kinds of policy approaches have the greatest chance of staying within the box so that the techniques chosen will survive legal scrutiny. To accomplish this goal, the chapter itself is divided into two main parts. The first part (Section 4.2) helps us understand the legal frameworks that apply in the coastal zone itself: What areas of law are impacted in coastal management generally, and what relationship do these laws share with one another. To provide this understanding, a background on the law is provided through the use of a *hierarchy of laws* conceptualization. The goal is to ensure that the reader has a basic understanding of how legal principles are related to one another. With a basic relationship among the different types of law established, the second part (Section 4.3) discusses traditional land-use planning laws that apply in the coastal zone. The emphasis in that section is on the Tenth Amendment powers reserved to the state (government), which include the power to regulate health and safety, and on the Fifth Amendment, which provides both a power to government (eminent domain) and a critical limitation on that power (the Takings Clause; Section 4.4). Collectively, these laws help us understand how the "box" is defined in coastal management generally, and more specifically how the "box" impacts sea level rise adaptation strategies discussed in this text.

Once a background of legal frameworks has been provided, the second goal of this chapter is to apply those laws to government activities in the coastal zone. To accomplish this goal, an in-depth look at government conduct will be done to distinguish the kinds of activities that fall within the box of lawful behaviors, and then distinguish them from activities that may fall outside the box. To begin, the question of *takings* under the law will be examined in some depth. A *taking* refers here to the unconstitutional dominion, physical occupation, or regulation of land by government that violates the Takings Clause of the Fifth Amendment to the United States Constitution. Much of the focus here will be on the *regulatory* takings issue because most of the adaptation strategies discussed in Chapter 3 of this text are created through government regulation that places limitations on how private landowners use their land.* Both the history of regulatory takings and its application to coastal management are discussed in depth in this section (Section 4.4).

The final section of this chapter (Section 4.5) discusses government options related to sea level rise adaptation based on the *role* government takes in planning and implementing policy directions. Many of the adaptation approaches mentioned in Chapter 3 require government to take the role of *regulator* in order to achieve its goal. However, government has other roles, including its role as *property owner* and also *trustee of public rights* in the coastal zone. Looking at these alternative roles of government as a way of avoiding regulatory takings claims will be analyzed as a means of developing and implementing policy goals for sea level rise adaptation in the coastal zone.

Again, the overall goal of this chapter is to provide a conceptual framework of what kinds of government actions fit clearly within the extent of legal powers (the box), while developing a sense of the kinds of strategies that help ensure that a government action does not go beyond its legal power. If coastal managers can master the legal frameworks that limit the kinds of actions allowed in adapting to sea level rise, then policy options can be created that meet the objectives defined elsewhere in this text while avoiding legal liability.

4.2 Background on Legal Considerations

As indicated in the introduction to this chapter, legal frameworks may be seen as a kind of "box" that establishes both powers and limitations for government actions. To place this dynamic in the context of public policy as a process of government conduct, a summary of government functions is necessary in order to understand how the types of actions undertaken by public officials in the coastal zone are understood. The mechanism employed for this purpose is a focus on the *hierarchy of laws* within the U.S. system of government. Through a discussion of the different categories of law that apply within this legal system, the roles of government will be identified to provide a greater understanding of how legal frameworks influence the functions of government (Rosenbloom, Kravchuk, and Clerkin 2008). This understanding will then form the basis for deciphering how government actions in the coastal zone are both empowered and limited by legal doctrine.

* It is generally obvious when a government action is aimed at using its eminent domain power to purchase the land. The land purchase options discussed in Chapter 3 are instances where government policy is aimed at paying just compensation for the property; this kind of policy response is **not** the focus of a regulatory taking. Rather, the focus here is on the various kinds of government regulations geared toward limiting the use of private property without removing ownership rights or when government physically occupies the property.

4.2.1 Hierarchy of Laws

The *hierarchy of laws* is conceptually presented in Figure 4.2. There are three main areas of law defined in the hierarchy: (1) *constitutional law,* (2) *statutory law*, and (3)*regulatory law*. Before getting into these three categories of law individually, they will first be related to the different *branches* of government that are *primarily* responsible for creating and implementing each area of law. Understanding the relationship between a branch of government and a category of law can provide insights into the law/policy dynamic being discussed in this chapter. For example, if the role of the legislative branch is to *make* law, then what role does the executive branch have in interpreting and carrying out the intent and purposes of that law? Can the executive branch make any interpretation of a legal standard established by the legislature, regardless of its connection to that legal standard? Or are there limits on executive interpretation, and if so, what are those limits and how are they determined? These are the kinds of questions that help place areas of law into a meaningful framework for understanding the different roles of government in relation to the law.

To begin, a general distinction between the branches of government and legal categories can be made. The *legislative* branch of government is generally responsible for making law and is therefore directly connected with most aspects of *statutory law*. The *executive* branch can propose policy directions and submit bills for vetting by the legislature, but the executive branch maintains a primary function of implementing and enforcing laws. The manner in which the executive branch often implements a statutory mandate through delegated powers granted in the statute is to create *regulatory law,* which allows for the development of regulations for the purpose of implementing legislative goals. The *judicial branch* of government is

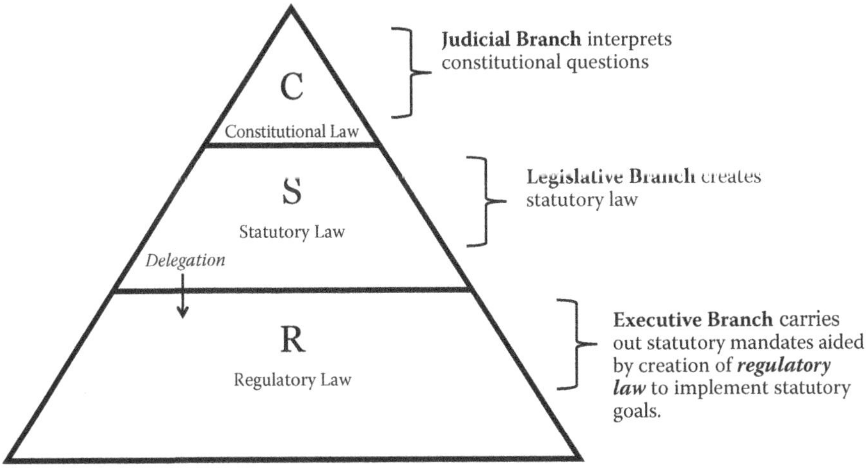

Figure 4.2 Hierarchy of laws representation.

the arbiter of legal doctrine. Beyond other important functions, the judical branch ensures that actions of the legislature and executive branches of government are done within limits set forth under *constitutional law* principles. The judiciary helps to define the meaning of the law in relation to constitutional principles, and by doing so helps to ensure that the actions of the legislative and executive branches are within the boundaries (the "box") of the law.*

The different responsibilities attached to each branch of government help place the role of the law into a perspective that relates to the powers and limitations of government. To understand the limitations the law can place on government actions, a greater explanation of the hierarchy between different areas of the law must first be explained. In Figure 4.2 the representation of the hierarchy is that of a pyramid, showing constitutional law at the top of the pyramid, legislative law at the center, and regulatory law at the bottom. This representation, from top to bottom, is meant to reflect a hierarchical relationship between these different kinds of law. The hierarchical nature suggests that constitutional laws are superior to both statutory and regulatory laws; and following this logic, statutory laws are superior to regulatory laws. The idea that an area of law is "superior" is really meant to reflect limitations that are placed on each category of law in relation to how the laws interact with one another. For example, regulatory law emanates mostly from the statutes, whereas responsibility for the implementation of statutory goals is delegated to an executive body. The regulations that are created from this delegation must be within the scope and purpose of the delegation itself; in a sense, the regulations must be within the boundaries created by the statute upon which the delegation is made. An example can help make this point more clear.

Consider the following: the U.S. Congress (legislature) passes a law (statute) that has a stated policy goal of cleaning the air; let us call this law The Clean Air Law. The Clean Air Law establishes a goal of clean air and identifies a general standard of air that is "clean enough to ensure the health, safety, and welfare of the citizens of the United States." Beyond this general statement, the law itself is silent as to what specific standards would equate to air that is "clean enough to ensure the health, safety, and welfare of the citizens of the United States." However, the law does delegate the responsibility for achieving this standard of clean air to the Federal Executive Branch, specifically the Environmental Agency within the executive branch. The Environmental Agency has authority under this delegation to develop "rules" (regulatory laws) to aid in implementation of The Clean Air Law

* There are a number of other functions carried out by each branch of government not expressly stated in this summary. The goal of the summary is to provide a direct relationship between each category of law in the hierarchy of laws to a government branch function. From this larger perspective, legislatures make law, executives enforce the law, and judiciaries interpret the law. Each branch may engage in some of these other activities as well (the executive engages in quasi-legislative and judicial functions when it is acting in its administrative function), but the role identified is meant to express the *primary* function of each branch.

goals. In addition, the Environmental Agency has its normal powers to enforce the rules as they represent the implementation of the statutory goal under the Clean Air Law. So the Environmental Agency goes about developing rules to implement The Clean Air Law goals. One of the rules established by the Environmental Agency regulates automobile engine design so those engines are no longer capable of emitting pollution into the air. The Environmental Agency creates this regulation even though The Clean Air Law specifically *excludes* automobile manufacturing from the goals and standards established under this law. The Environmental Agency also decides to regulate minorities, passing a regulation to implement The Clean Air Law that prevents minorities from driving automobiles because, according to the Environmental Agency, minorities tend to have older cars that emit more pollution. Thus, by preventing minorities from being able to drive, the goals of The Clean Air Law will be obtained. There is nothing explicitly written in The Clean Air Law itself that limits the ability of the Environmental Agency to regulate minorities as a means of achieving the goal of clean air.

Now that the example has been provided, let us take some time to parcel out the issues that arise in relation to our hierarchy of laws analysis. Remember that the goal here is to see how the constitution, statutes, and regulations interact with one another in the hierarchical scope described above. The first analytical step is to decipher the legal relationships between the laws and regulations created. This can begin with a review of The Clean Air Law itself, an example of *statutory law*. As a statute, the law sits below constitutional constraints and above regulations related to the statute, indicating that the regulations that evolve from the statute must be in accordance with the powers delegated by The Clean Air Law. Conceptually, we can think of this relationship in a set of boxes where the U.S. Constitution forms the outer box, The Clean Air Law sits within the box created by the U.S. Constitution, and the regulations created by the Environmental Agency sit within the box that encompasses The Clean Air Law. The relationship established indicates that The Clean Air Law must be in accordance with federal constitutional principles, the regulations enacted by the Environmental Agency must be in accordance with the powers delegated to it under The Clean Air Law, and these regulations must also be in accordance with federal constitutional principles. Acknowledging these legal relationships and limitations, the legal issues presented can be analyzed.

Beginning with The Clean Air Law, the purpose and intent of the law can be analyzed to determine the relationships between the statute and the Environmental Agency's actions in relation to the statute. The Clean Air Law expressly delegates authority to the Environmental Agency for the development of regulations that aid in the implementation of the statute's goals; in this way, we see that the Environmental Agency has authority to engage in the creation, enforcement, and interpretation of regulations to implement the statute's goals.* However, the question here is not whether the Environmental Agency has *general* powers to enact regulations to implement the statutory goals under the Act, but whether the actual regulations developed by the Agency are *specifically* within the powers granted to the Agency by the statute. To answer this question, we need to look at the specific regulations developed by the Agency.

The first rule mentioned in the example above indicates that the Environmental Agency is attempting to regulate automobile engine design even though The Clean Air Law specifically exempts automobile manufacturing from the law. The question then is whether the regulation is outside the powers granted to the Agency under the statute. Or, said another way, does the Agency rule regulating the design of automobiles fall outside the "box" of authority created by the statute? Figure 4.3 represents this question visually.

The two "Xs" placed within and outside the "Statutory Delegation" box indicate different answers to this question. For example, the X contained within the "Statutory Delegation" box suggests that the Environmental Agency was delegated the power to enact the regulation. The X that is outside the "Statutory Delegation" box indicates that the Agency does not have authority to enact the regulation and the regulation is of the kind that is outside the scope of the statute itself. Based on the facts, it is likely that the rule developed by the Agency here to regulate automobile

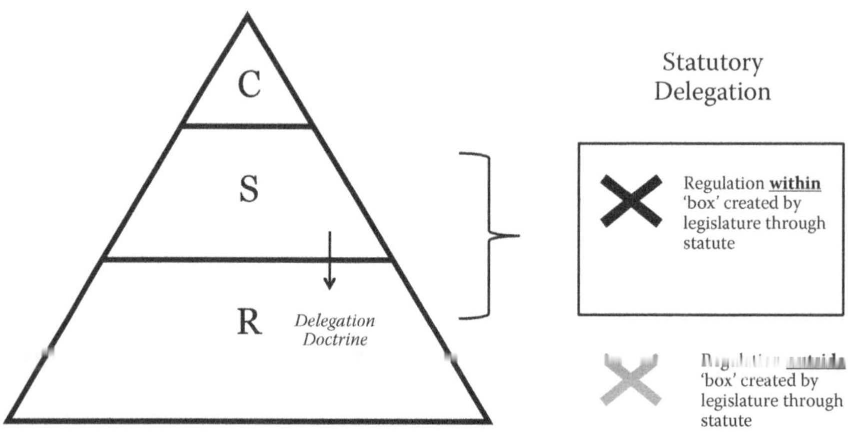

Figure 4.3 Representation of Agency Regulatory Authority.

* The ability of the executive branch of government to engage in rulemaking, enforcement, and review of the rule represents powers that have been acknowledged to exist in administrative entities under principles of *administrative law* adopted at both the federal and state levels of government.

engine design represents the X that is outside the "Statutory Delegation;" the plain language in The Clean Air Law clearly exempted automobile manufacturing from the purposes of the statute. Thus, the Agency's attempt to regulate automobile engine design *based on The Clean Air Law* is unlawful.

The second rule that is found in the example above is a regulation that prevents minorities from driving automobiles. There is no express prohibition of this kind of regulation in The Clean Air Law itself (the statute prevents direct regulation of automobile *manufacturers*, but is otherwise *silent* as to the regulation of those who *drive* automobiles). So the question presented here from a legal analysis standpoint is twofold: First, is the regulation a reasonable interpretation of the power granted to the executive agency by the legislature in passing the statute and thus within the scope of powers granted under the statute? And second, if the regulation is deemed to be within the scope of powers granted to the agency under the statute, is there any other legal principle that limits the regulation beyond The Clean Air Law itself?

The first question—whether preventing minorities from driving is a reasonable interpretation of the powers delegated to the executive agency by the statute—is a way of interpreting the scope of delegation authority. The question is really one of judicial interpretation, where the kinds of analytical tools that judges would use to understand the relationship between the statute and the regulation are being utilized. In this case, the relationship between the statute and the regulation is not *expressly* clear, meaning there is no provision in the statute that expressly directs the agency to regulate minorities in driving automobiles to achieve the goals of cleaning the air. When there is no express provision, the next step in judicial interpretation is to look at the *purpose* of the statute and to see if the regulation reasonably meets that purpose. In this case, the purpose of the statute is to clean and protect the quality of the air. The purpose of the regulation seems aimed at limiting air pollution because it is based on preventing minorities from driving, and preventing minorities from driving is said to be linked to evidence that minorities drive older vehicles that pollute more than non-minorities; whether there is "evidence" to support this proposition provides the basis for determining whether the regulation is within the scope of the statute; some objective hard evidence would likely need to be shown to substantiate this proposition as fact. In addition, it is likely that some evidence would be required to show that the class of citizens (minorities) disproportionately uses older vehicles in comparison to all other citizens. In short, there would have to be support from the agency to show a connection between the proposed regulation and the goals established under The Clean Air Law. Thus, in summary, the regulation may be shown to directly apply to a delegated responsibility from the legislature where the regulation follows an *express* provision of the statute, or the regulation can be connected to the statute by showing that it reasonably carries out the purposes of the statute.

The second question brings us back to our hierarchy of laws analysis. Assuming that the regulation of minority drivers is found valid under The Clean Air Law, the question becomes whether the regulation is *invalid* under any other law(s) that are above the regulation; this question can be answered by looking at other statutes that might be impacted by the regulation, as well as looking at federal constitutional provisions that may be violated by the regulation. It should be obvious that the regulation of minorities, including the assumptions behind the regulation, is highly *discriminatory*! As such, the regulation likely violates other federal laws that prohibit discrimination. When two laws of equal prominence are at-issue (federal statute to federal statute in this example), the courts will generally harmonize those laws to ensure that no conflict exists between them. In this case, the interpretation of the regulation under The Clean Air Law might show no conflict; however, the regulation is likely in conflict with prohibitions against discrimination contained in other federal laws. In this case, the regulation cannot be harmonized with those other federal laws (by enforcing the regulation, the Environmental Agency would be violating the tenets of other federal laws). Thus, the regulation is likely *invalid* for this reason alone. In addition, the regulation is certainly discriminatory against a class of people based on socio-economic and other factors. This classification creates a disparate impact on this class of people (not being allowed to drive automobiles) and is thus very likely a violation of federal constitutional law. Because the constitution sits atop our hierarchy of laws, the regulation will likely be found invalid under constitutional principles.*

The example above shows some of the major hierarchy of law principles in action. The overall validity of The Clean Air Law depends on a variety of factors. For example, an express provision in the statute that is deemed unconstitutional would be stricken down because the constitution trumps the statute; if the provision could not be stricken from the statute without the entire purpose of the statute becoming implicated, then the entire statute itself may also be stricken as unconstitutional. Further, even valid statutes can delegate responsibilities to executive bodies in ways that result in violations of the law. Executive agencies can develop regulations that are outside the delegation of powers granted to it by the statute, as is the case when an agency creates a regulation that is unconstitutional (it is presumed that legislatures do not delegate a power to an executive

* If the regulation were a result of an *express* provision in The Clean Air Law itself (if the statute included a provision that indicated minorities are to be prevented from driving automobiles), then The Clean Air Law itself would be found unconstitutional. The effect would be to either strike down that provision of the statute dealing with minority ownership of automobiles, or if that provision could not be separated from the statute without destroying the purpose of the statute itself (which does not seem to be the case here), then the entire statute would likely be found unconstitutional and thus invalid. In judicial review, the courts seek to "read" a statute as having a lawful and constitutional purpose as much as possible under the presumption that the legislature passes constitutional and otherwise lawful pieces of legislation. Thus, "carving out" unconstitutional provisions is generally the preferred route of judicial review rather than striking down an entire statute. This preference reflects deference from the judiciary on the separation of powers doctrine by noting the legislature's primary role in enacting legislation.

branch to engage in unconstitutional acts to implement legislative goals). Any executive agency action that violates a constitutional tenet is automatically presumed to be outside the delegated authority of the legislature; where the agency action is unconstitutional but within an express provision of the statute, both the regulation and the statute (or statutory provision) will be found invalid under the constitution. Through analyzing this interrelationship between constitutional law, statutory law, and regulatory law, a better understanding of the hierarchy of laws is presented and ready for application to coastal zone management and issues surrounding adaptation strategies to sea level rise.

So just how does an understanding of the hierarchy of laws described above help decipher the legal considerations that apply to sea level rise adaptation policy? For one thing, the relationship between the different branches of government and the law establishes a foundation for understanding the potential interactions between adaptation policy development and the legal frameworks discussed later in this chapter. Consider that most coastal land-use planning occurs as an administrative function within the executive branch of government at the local level. This means that many of the operative aspects of policy development and implementation will occur at either the legislative or regulatory law levels, with regulatory laws representing the bulk of the "rules" that are developed to implement policy directions. Consider the various adaptation approaches and techniques discussed in Chapter 3; the vast majority of those approaches would be developed by local executive entities, or through a process where local legislative authority (ordinances and bylaws) is used to empower executive rule-making functions. In either case, the "source" of the government power being utilized to develop adaptation approaches will be subject to the kinds of limitations described in the example provided above. For example, the development of a zoning overlay district to implement a proactive adaptation approach that includes buffer zones and minimum setback requirements may be challenged on a number of legal grounds. To assess the validity of these challenges, the basis for the zoning overlay district will need to be identified; is the overlay district being created through a "legislative" action or through a "regulatory" action that is implementing a power to zone granted by a previous legislative enactment? The answer to this question will then establish the basis for analyzing the legal challenge(s). If the basis of the overlay district is regulatory in its inception, then the analysis must include a review of the power granted to the agency (is there power to create the overlay district?). In addition, any legal challenge based in statutory authority may overrule the regulatory act (because statutes sit above regulations under the hierarchy of laws). If the basis of the legal challenge emanates from the U.S. Constitution, then the zoning district (whether emanating from a statute or regulatory act) is subject to being overruled as a violation of constitutional protections.*

Understanding the hierarchical nature of several categories of law can help in assessing the potential impact that legal frameworks may have on a policy direction. In addition, understanding these legal relationships can aid in policy planning; a well-thought-out policy goal can be better achieved if the legal implications are considered part of the planning process (Rosenbloom, Kravchuk, and Clerkin 2008). Regulations can be adopted that are more in line with statutory requirements. Statutes can be enacted, if need be, to ensure there is an adequate legal basis for moving in a particular policy direction. Most importantly, both statutes and regulations sought to implement sea level rise adaptation goals can be reviewed to ensure that they are as compatible as possible with constitutional principles. These kinds of considerations highlight the remainder of this chapter by focusing on two important questions:

1. How do existing legal frameworks impact coastal adaptation planning?
2. What strategic options exist for government in seeking to mitigate the impact of legal frameworks on coastal adaptation planning?

The first question is explored by reviewing traditional land-use planning and the role of the takings doctrine in relation to land-use planning. The second question, while not definitively answered, is explored in the final section of this chapter where the nonregulatory role of government is highlighted as a way of steering clear of legal framework issues in policy implementation, particularly constitutional takings doctrine. The answer to the first question examining existing legal frameworks' impact on coastal adaptation policy begins with a review of land-use planning law.

4.3 Traditional Land-Use Planning Laws

Land-use planning is fundamentally a local government operation, where most land-use decisions are made at the municipal level. However, traditional land-use powers emanate from the highest source of law, the U.S. Constitution. Under the Tenth Amendment to the U.S. Constitution, all powers not specifically given to the federal government are reserved to the states (Leuchtenberg 2002). This includes the power to enact laws and otherwise regulate for the health, safety, and welfare of the states' citizenry. Included under these *police powers* is the ability to regulate land use. Although states have the power to regulate land use within their borders,

* Constitutional claims are the basis for *takings* claims (violations of the Fifth Amendment to the U.S. Constitution) and thus tend to be some of the more popular challenges to coastal land-use restrictions because a showing of a constitutional violation will invalidate both statutory and regulatory law.

that power is not unlimited. The ability to regulate land is limited in a number of ways, but conceptually for the focus intended in this chapter, the major limitation placed on state regulatory authority derives from the Fifth Amendment to the U.S. Constitution, specifically the takings limitation as expressed and interpreted under that amendment. The interplay between the Fifth and Tenth Amendments to the U.S. Constitution helps establish an important starting point for understanding how legal frameworks impact the development and implementation of coastal management policies geared toward adapting to sea level rise. The purpose of this section is to discuss the Tenth and Fifth Amendment powers and limitations to provide a baseline of government power so that actions that evolve from these base powers can be better understood in context. However, prior to engaging in an exploration of each amendment in detail, a short summary of property law principles affecting the coastal zone is provided to gain an overview of the subject.

4.3.1 Property Law Principles That Apply to the Coastal Zone

Major property law principles that apply to the coastal zone include a confluence of *common law* (sometimes "codified" by statute), *constitutional law*, and *general zoning powers* of state and local governments (Ackerman 1978). *Common law* is a term that refers to concepts derived from a long history of unwritten but enforced legal principles (Holmes 1909). Examples that apply in the coastal zone include *the public trust doctrine*, a common law principle that states that "submerged lands" (land regularly inundated by water) are "public lands" that are owned by the public and held in trust by the government for the benefit of the public. The public trust doctrine has a tradition that evolved as long ago as The Roman Empire, being adopted by England and ultimately the United States (Archer 1994). Other common law traditions include *custom* where customary use of land is given legal significance over time, and *nuisance* where a private landowner can stop activities that are shown to cause an unreasonable interference with the use and enjoyment of the private landowner's property (Caldwell and Segall 2007, p. 552–55).

These common law traditions that apply to real property are often referred to as "background principles of property law" because they help establish the conditions upon which property rights exist. For example, under the public trust doctrine, one who purchases private property that abuts the ocean accepts ownership of that property under a preexisting condition, the condition being that the property rights of that private landowner are subject to change as the sea level changes. Where sea level moves inland, the private landowner loses some property; where sea level moves seaward (e.g., from the slow natural addition of sand, often referred to as *accretion*), the private landowner gains additional land (Sax 2010). The private landowner cannot complain about the possibility of losing some private land to sea level rise because the right of public ownership in submerged land is a "background principle of law" that limits the rights and expectations of the adjacent private landowner. This same argument can be made of the common law prohibition against *nuisance*; private landowners take land subject to the limitation that they cannot create a nuisance on their property, particularly the kind of nuisance that threatens the public (Caldwell and Segall 2007, p. 557–558). Although this prohibition against public nuisance may not exist in local or state statutes (i.e., the concept may not be *codified*), and although this prohibition may not be explicitly stated in the contract or deed for the sale of the land, the prohibition is imbued into the use of the property as a "background principle of property law." In essence, the owner accepts the land with these limitations already in existence as a matter of law—they exist in the "background," and real estate bought and sold is subject to these background conditions.

Collectively, common law traditions impact coastal management in a variety of ways. More specifically, they help to formulate expectations between private and public rights in the coastal zone. The public trust doctrine identifies the public's ownership of submerged lands (codified federally under the Submerged Lands Act of 1953, 43 U.S.C. §1301 *et seq.*). Doctrines of nuisance and custom also shape the expectations of coastal property owners, supporting limitations on how private coastal property is used. As such, common law traditions are important aspects of property law principles that apply in the coastal zone because they help manage expectations on how property is used under background principles of law. Alternatively, *constitutional law* principles can be seen as a *reflection* of the expectations of public and private interests in coastal land. Where common law traditions help to manage expectations in the land, constitutional principles help to define those expectations. To better understand how these legal frameworks help manage expectations in private land use, some exploration of the principles from which private property expectations are derived is warranted.

There is an underlying principle that impacts property right expectations in land defined through the following inquiry: whether property rights are *"natural"* and therefore preexist political institutions (Ely 2008, 703), or whether property rights are solely the result of *social constructs* such as government (Bentham 1840, p. 137–39). A representation of this query is shown in Figure 4.4 and Figure 4.5.

In Figure 4.4, the "natural" property rights theory is represented by showing social institutions existing within a natural property right framework; the social (and thus government) functions are subject to the natural property rights that exist outside the social institution. Figure 4.5 shows the opposite, wherein the property right sits within the social institution from which the right is created. Whether a property right sits outside a social institution or is wholly defined by social institutions determines in large measure the extent to which property rights are subject to regulation.

If property rights are "natural" rights that preexist social institutions, then a strong argument can be made that suggests that changes in social expectations over time cannot reduce these natural rights without impinging on property rights. Those who follow this ideology tend to find the Fifth Amendment takings prohibition as a measure to ensure that fundamental property rights survive changes in

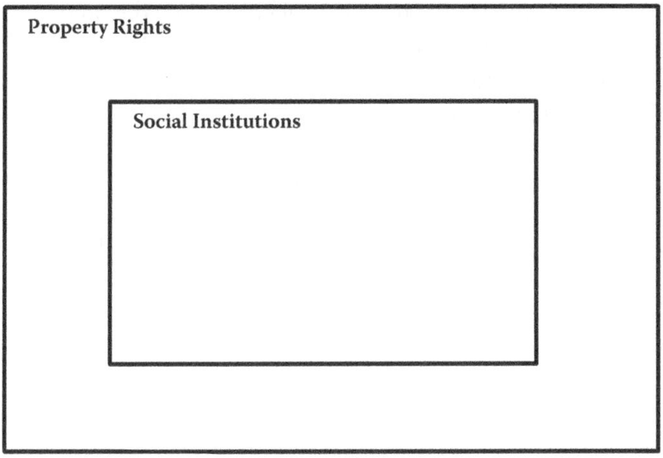

Figure 4.4 Representation of *natural* property rights theory.

societal expectations. Thus, any government activity that impacts these rights to the point where they are effectively removed cannot occur without triggering eminent domain proceedings. In essence, government conduct is restricted in regulating private property to the extent that the government conduct removes the property right; the further government action goes toward limiting a property right, the more likely it is that the government has engaged in an unlawful action.

If property rights are created through a social construct and thus flow directly from political institutions, then those rights should be logically subjected to greater restrictions by government without resulting in a requirement that government pay for the restriction even where the restriction effectively removes the property right

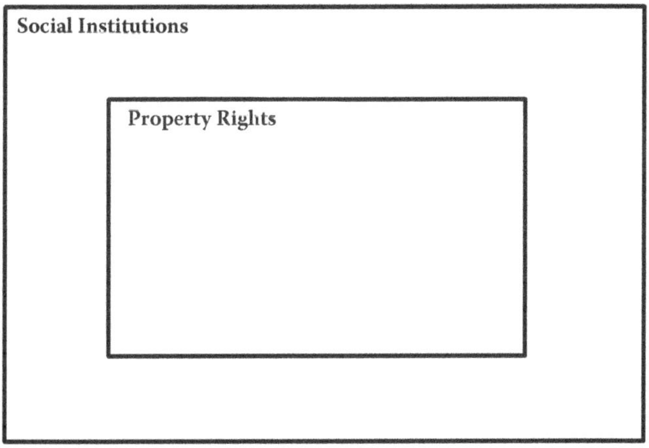

Figure 4.5 Representation of *social construction* of property rights theory.

in question. In this ideological stance on property rights, the police powers of the state are seen as superior to private property rights; and when there is a clear connection between the government conduct and the health, safety, or welfare of the citizenry, the government action will be upheld even where private property rights are reduced in the process.

From a constitutional law standpoint, this fundamental argument is linked to the question of how the Tenth and Fifth Amendments to the U.S. Constitution are interpreted when there is the need to regulate private land under a legitimate police power of the state. To understand where the line is being drawn, the judicial branch of government is required to interpret the law as applied to different factual settings. The U.S. Supreme Court has provided a number of decisions in this area to help in understanding when the exercise of government power might cross a line that leads to a finding that a private property right has been unlawfully inhibited, if at all. Central to this analysis is the underlying philosophical stance as to where private property rights emanate from and thus the relationship of government actions to those rights.

Finally, *general zoning powers* are the fundamental "tool" employed by government to regulate land (Ackerman 1978). The police powers of the state grant government the ability to restrict land without seeking title to the property.* Zoning is the fundamental process by which government engages in its regulation of land, and it has important application to coastal areas. Recall that one of the primary policy techniques described in Chapter 3 was the development of zoning overlay districts. These districts are developed and implemented under the general zoning power of government and reflect the inherent planning that goes into a zoning process. The capacity to zone for different uses of land based on the land's relationship to physical characteristics (e.g., land abutting the coastline) and different kinds of uses of land (e.g., separating residential uses from industrial uses) provides the government with the ability to plan for the safety and welfare of its citizenry. The extent to which zoning regulations can be incorporated into existing land uses is the question that brings in the earlier discussion about the formation and evolution of private property rights. As noted in the zoning overlay district example, changing expectations on how land is used may include altering existing uses. When existing uses are altered through government action, the potential for claiming a taking of individual rights exists. In reviewing the takings question, the public and private rights at issue must be interpreted and judged. How that judgment ultimately concludes is based on a number of factors, with the police power of the government to zone as a central factor underlying support for the right to regulate coastal areas.

---

* The power of government to regulate the use of land was upheld by the U.S. Supreme Court in *Euclid v. Ambler Realty Company*, 272 U.S. 365 (1926), where "reasonable" government zoning regulation powers that placed limitations on private property for the benefit of the public at large were upheld.

Land-use planning emanates from a number of property law principles. Common law traditions help establish the various expectations that are held between public and private interests in land. For example, private property rights give way to public rights in submerged lands in the coastal zone; if submerged lands move landward due to sea level rise over time, then the private upland becomes public submerged land by operation of law. No private landowner can expect the newly submerged land to remain their private property because the doctrine of law preexists their use of the land. Similarly, zoning laws that allow for public control of private land also help manage expectations between public and private interactions. The extent to which public control of land is allowable depends in large part on from where the power to control land emanates. From a constitutional law standpoint, the Tenth Amendment acknowledges the police power of the state to regulate land, although the extent to which land can be regulated is bounded by the Fifth Amendment takings limitation. Which property right principle is superior under a given circumstance, the police power of government to regulate or the private property right being regulated? This question cannot be conclusively answered, but the answer for a particular factual setting is based on the evolution of the property right principles discussed above. Where property rights evolve outside of social institutions, more power is given to the property right. Where property rights evolve from social institutions, the property right is subject to greater control. This dichotomy between public and private rights will help underpin the issues discussed in the remainder of this chapter.

4.3.2 Interactions between Tenth and Fifth Amendment Powers and Limitations

With a summary of the property law principles that apply to the coastal zone completed, the focus now turns to understanding the interactions that occur between the Tenth and Fifth Amendment powers, and how these powers create a set of legal standards that apply to coastal zone policy in general and sea level rise adaptation strategies in particular. Recall that the Tenth Amendment is an acknowledgment in the U.S. Constitution of the reserved powers of the state, which include the power to regulate for the health, safety, and welfare of the state's citizenry (Leuchtenberg 2002). The Fifth Amendment, particularly the takings limitation of that amendment, places a limit on the police power of the state, identifying a "reasonable" standard limitation on the police power to regulate private property rights in land; where a police power goes too far and is deemed "unreasonable" in its regulation of land use, the Fifth Amendment requires the government to pay compensation for the limitation placed on the private property right (Lazarus 2008; McGuire and Hill 2012). This interplay between constitutional powers is displayed as a continuum in Figure 4.6.

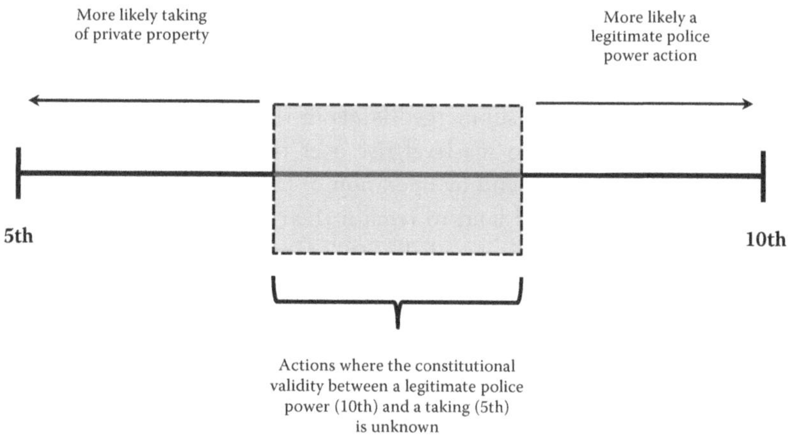

Figure 4.6 Continuum of Fifth and Tenth Amendment powers.

Some point on the spectrum between Tenth Amendment powers and Fifth Amendment limitations is the upper-limit where government actions in regulating land use are no longer seen as reasonable and thus wholly within its reserved powers. As shown in Figure 4.6, when this point is crossed, the government action becomes unreasonable and thus triggers Fifth Amendment takings analysis. Just what constitutes this point is often a matter of judicial interpretation and the takings cases that have evolved to help define this point are examined in greater detail later in this chapter.

The goal in this section is to provide an overview of Tenth and Fifth Amendment interactions to identify the impact these interactions have on developing legal standards in coastal zone management. There are three main points to highlight in this section.

1. The acknowledgment that these interactions are occurring at a level of *constitutional law*, and thus at the highest level of the hierarchy of laws described at the beginning of this chapter; the implications for coastal development in bounding lower legal "hierarchy" actions must be acknowledged.*
2. The impact of uncertainty as to the extent of government interactions that may cross over from reasonable to unreasonable "regulation" of private property rights must be acknowledged and explored; understanding the impact of uncertainty in this area is important when judging policy options for sea level rise adaptation.

* For example, any coastal policy development done at the regulatory stage that violates a constitutional principle, even if presumptively allowed by statutory delegation, is unconstitutional and thus void as a matter of law.

3. The foundation upon which a Tenth Amendment police power is based—the role of government as "regulator"—must be identified and explained so that alternative options outside of a regulatory stance may be identified and explored when thinking about policy directions that may avoid takings challenges.

4.3.2.1 Constitutional Nature of Tenth and Fifth Amendment Interactions

Any discussion of land-use planning that involves the Fifth and Tenth Amendments of the U.S. Constitution brings the hierarchy of laws principles to the forefront. If the source of power that is justifying a government action derives from constitutional law, then those constitutional standards provide a base level of protection against which all actions must be judged, whether those actions emanate from a statutory or regulatory law source. This means that policies aimed at regulating expectations along coastal areas having a direct impact on private land use will lead to a legal analysis that ultimately judges the constitutional validity of the policy direction. Because both the power to regulate land and the limitations on that public power emanate from a constitutional source, the basis for prohibiting the private exercise of property rights must be closely examined within a constitutional context.

There are a few points that derive from the constitutional nature of land-use regulations as described above. First, any government action that attempts to regulate land use in the coastal zone, whether directly or indirectly, is subject to a constitutional analysis balancing the police powers utilized to achieve a public goal against limits on those powers as defined by the U.S. Supreme Court when private property rights are unconstitutionally constrained. Second, when engaging in policy directions that impact land use, particularly the private property rights of landowners, the policy must be reviewed to determine the likelihood that the limitations placed on private land use may be challenged on constitutional grounds. If there is a likelihood of such a challenge, policy planning must include an assessment of the potential risk for the challenge balanced against the capacity of the government entity to absorb the risk.* Third, the impact of uncertainty must be considered in policy planning and development stages; where a policy direction contains greater uncertainty of the legal consequences of a policy direction, there should be some acknowledgment of this fact. Alternative approaches that reduce legal uncertainty, particularly constitutional takings claims, should be identified and balanced against the uncertainty of risk posed.

* For example, if the risk includes the finding of a "taking" of private property rights without just compensation, then the capacity to absorb this risk by the government should include the ability to pay just compensation to the affected landowner(s) to carry out the policy direction.

4.3.2.2 Impact of Uncertainty in Government Regulating Coastal Land Use

Uncertainty attaches to government regulation of land use in the coastal zone because of a lack of clarity in understanding when government conduct might move from a reasonable regulation to an unconstitutional taking as visually represented in Figure 4.6. The basis for this uncertainty derives from the fact that case law is the main mechanism for deciphering when government regulation of land goes too far, and that case law, while helpful, is incomplete when applied to different factual settings.* The uncertainty also stems from the fact that the public and private rights expressed through land-use regulation *both* emanate from constitutional powers; the private right to be free from public actions that unreasonably deprive rights in land is no more powerful than the public right to place reasonable controls on the use of land. It is not a hierarchy of laws situation that determines legal rights and wrongs here, but rather the mix of government interaction with private rights in the land. The uncertainty, then, does not stem from the legal frameworks from which the respective private and public powers derive, but rather the manner in which the interactions between private and public rights can result in either a valid expression of public right or an unconstitutional infringement on a private right.

Uncertainty can be mitigated in several ways. Obviously, government can decide to focus on the *public purpose* aspect of their actions that impede private rights and otherwise be prepared and willing to pay compensation for policy directives shown to "take" those private rights in land.† If government is not willing to pay for the privilege of protecting coastal citizens from sea level rise, then it must look to reduce the uncertainty that exists when it acts under its police power authority. Government

* The major U.S. Supreme Court cases that help interpret regulatory takings claims against government conduct are outlined in Section 4.4 of this chapter.

† Government can always act under its *eminent domain* power as outlined in the Fifth Amendment to the U.S. Constitution. Under this power, government has the capacity to "take" private land where it does so for (1) a public purpose; and (2) where just compensation is paid to the private landowner. Government can always adopt a policy approach that looks to regulate land uses within the coastal zone and simply "take" the private rights under its eminent domain power and thus pay for the taking. An example might be the use of a buffer zone technique as described in Chapter 3 that prohibits development between the zone and the shoreline. The government may simply appropriate strips of private land between the shoreline and the upland area to create the buffer zone and pay private landowners for the appropriated land. If the buffer zone is done to ensure the safety of the citizenry, then it will likely be seen as a valid *public purpose.* Thus, all the government must do is pay fair-market value for the taken property rights of private landowners in order to finalize implementation of the buffer zone strategy; there are no constitutional issues to consider when the public purpose is proven and the government has paid just compensation. If government does not wish (or cannot afford) to pay for the buffer zone but desires the zone as a means of implementing important public protections, then the reasonableness of the restriction will be judged in light of the impact the restriction has on affected private property owner rights, bringing the discussion back to a constitutional tug-of-war between private and public interests in the land.

has capacities and roles that exist outside that of a *regulator* of land. Indeed, government, like private landowners, is also an *owner* of public land and can advance its ownership rights in a way that is not regulatory in nature (Sax 2010; McGuire and Hill 2012). In addition, government has obligations as *trustee* of the public's rights in both public and private land. Because these trustee obligations are nondiscretionary in most cases, government can enforce the public's rights in coastal resources without developing a "regulatory" stance and therefore help to avoid the constitutional law questions that arise when government *regulation* impedes private property rights in land (Caldwell and Segall 2007; McGuire and Hill 2012). These nonregulatory roles of government are explained in greater depth in Section 4.5.2 of this chapter.

4.3.2.3 Role of Government as Regulator under the Tenth Amendment

As indicated immediately above, government has roles that exist outside that of its *regulatory* stance when it comes to creating policy options about how it achieves certain goals in the coastal zone. The final point to be made in this section on traditional land-use planning laws is that government becomes a *regulator* of private property rights when it engages in traditional land-use management. When this happens, the Tenth and Fifth Amendment issues described above emerge and constitutional issues arise. This point is important because government often engages in passing statutes that are meant to implement policy directives. The statutes are often passed under the police power of the state but left to executive agencies to implement. It is at this point that implementation of legislative intent becomes an issue; if the implementation actions taken by local administrative bodies are seen as unreasonable, then there is a chance the actions may be subject to judicial review as an unconstitutional deprivation of private property rights. Thus, the traditional role of regulation, even when properly created under statutory powers, may lead to unconstitutional implementation through delegation to local executive bodies. Government must be aware of its role as regulator when implementing land-use policy because it has impacts on the property right expectations between private and public interests.

Traditional land-use planning laws combine a mixture of common law, constitutional law, statutory law, and regulatory law mechanisms. At the highest level, land-use planning and regulation is a function of the police power of the states confirmed through the Tenth Amendment to the U.S. Constitution; all representations of regulatory law that emanates from government as a means of controlling land use is supported by the Tenth Amendment. Limitations on the police power of the state to regulate land use exist primarily in the Fifth Amendment's prohibition against the taking of land without just compensation. Generally, a taking occurs when the government regulation of private property rights is deemed unreasonable. However, the question of what constitutes unreasonable government interference with private property rights is not entirely clear. Case law mainly in the form of U.S. Supreme Court opinions gives some guidance on when a taking may be found.

Exploring the background of takings jurisprudence generally, and the history of regulatory takings cases in particular, can help provide further understanding of the impact of these legal frameworks when government is approaching adaptation policy through techniques that regulate private land use.

4.4 The Takings Question

Takings law highlights a fundamental concept of American property law: Private land ownership is a foundation upon which individual freedoms and economic interests are based (Ward 2010). It is argued that removing the rights attached to land ownership fundamentally undercuts this principle and can lead to an imbalance between the individual citizen and government power protected through the first ten amendments to the U.S. Constitution, *The Bill of Rights* (Amar 1998). So fundamental is this concept of property rights protection that the line between a valid and necessary exercise of government authority and an invalid impact on private property rights is to some degree inexact and unknown. The purpose of this section is to review the history of takings law, including its evolution from controlling overt actions by government to physically "taking" private property to less-obvious "regulatory" takings claims where the government's act of regulating land is said to result in an effective taking of private property rights. The goal is to provide the reader with a deeper understanding of the concepts involved in takings law because of the impact this particular set of legal frameworks has on coastal policy approaches. Because many of the policy techniques described in Chapter 3 include some form of regulatory stance by government in controlling coastal land-use expectations, an examination of takings law is necessary to create a context for understanding when a threshold between a reasonable government expression of police powers (Tenth Amendment) crosses over into an unreasonable interference of private property rights (Fifth Amendment). With this knowledge in hand, coastal planners can better anticipate the kinds of actions that might trigger a takings claim, and therefore how to mitigate this potential in the planning and execution of coastal land-use restrictions to accomplish sea level rise adaptation goals.

4.4.1 Background

To understand takings law in general, it is important to begin by reviewing the historical antecedents associated with land-use rights in the United States. The history of land-use and private property rights helps to place the question of takings law into context, thus allowing the rationale of judicial decisions approaching takings analysis to be placed into perspective. It is this perspective that is key to properly utilizing takings law frameworks as a way of orienting policy planning because, as noted above, a takings law precedent is very much dependent on case law analysis; the basis for takings claims derives from constitutional power sources: Tenth Amendment police powers supporting the public power to regulate land use, and

Fifth Amendment takings prohibitions against public actions that unreasonably interfere with private property rights. Because both powers emanate from a constitutional law source, there is no way for statutory or regulatory laws to develop that violate these constitutional rights. Thus, judicial decisions through case law are the means by which we understand when a government act goes too far and results in a taking of private property rights.

An overview of takings doctrine begins with an understanding of property law principles that apply to land generally, and the special relationship these principles have in coastal areas. As noted in the previous section, interactions between police powers of the state and limitations placed on the exercise of those powers when applied to private property rights in land emanate from two competing principles of property law (see Figures 4.4 and 4.5). Where property rights are deemed to exist *outside* social institutions (including government), then courts as reviewers of the law will tend to grant government less latitude in restricting property rights. Alternatively, where property rights are seen as existing *within* social institutions, courts will tend to grant government more latitude in restricting property rights (Anderson and Hill 2003). In this way, the underlying philosophy of the justices deciding these cases becomes important because there is no agreement or mandate on where property rights exist in relation to social institutions. Judges who subscribe to the ideology that property rights *preexist* government will tend to favor a restrictive view on interfering with property rights. Judges who believe property rights are wholly *created* through government and thus flow from social institutions will tend to favor a more liberal view of government's ability to regulate property rights.* Therefore, the makeup of a judicial panel reviewing the government action, including their individual ideologies on the creation of property rights (something that is often not known, and if known, is not often acknowledged publicly) becomes an important consideration in understanding takings doctrine.

The creation of property rights is one factor in understanding takings doctrine. A second factor of equal importance includes understanding how property rights evolve at the coastal zone, an area of special consideration because of its unique attributes where property lines are often ephemeral and changing due to the dynamic characteristics found along the coastline. Coastal states (states that border large bodies of water) each have a unique history of how property rights are created

* Consider this statement from the perspective of how to define a *reasonable* use of government's police power to regulate private property rights in land. If the rights to land flow from the creation of government institutions, then those rights are subordinate to the right of government to protect the health, safety, and welfare of its citizenry. Thus, the emphasis on reviewing government action that limits private property rights will be on the *nature* of the government act. Once the purpose of the government action is found to be within its police power rights, then it must be assumed that the private property right gives way to the government action because the property right was created within the construct of government functions for the social welfare of its citizenry. It is much easier to support government actions where the property rights are deemed to derive from social institutions and not outside those institutions.

and defined within their respective coastal zones (Craig 2007; Craig 2010). The history of each state in how it defines its property rights, as "background principles" of state property law, is the basis upon which government regulations are often judged in relation to their impact on private property right expectations. The reason for this is that the private property owner's expectations in the kinds of property rights that exist depend in large part on the preexisting property rights, including historical limitations on those rights, of a particular area. One example to highlight this difference in the coastal zone is where the line of demarcation between private and public rights exists. A few coastal states[*] maintain that private property rights exist up to the low mean (average) tidemark, and thus private property owners abutting the coastline in these states own from the point of their upland private property down to the low tidemark. Most coastal states hold that private property rights end at the *high* mean tidemark. In these states, private ownership ends at the high tidemark and public rights include the intertidal zone. The difference between these two property rights designations is visually displayed for reference in Figure 4.7.

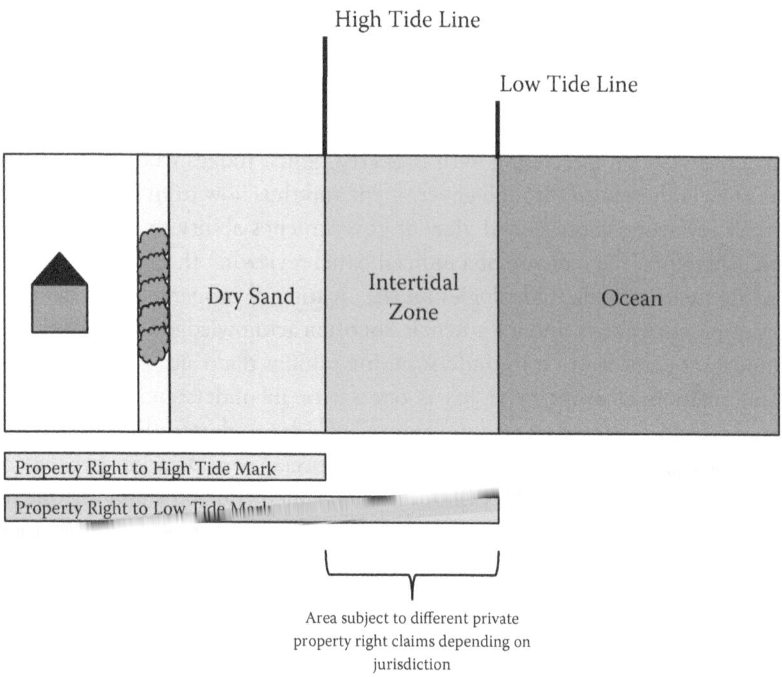

Figure 4.7 Representation of coastal property rights demarcation: public and private.

[*] Massachusetts and Maine are two examples of states that enforce private property rights to the low mean water line.

If a government were to regulate the intertidal zone in a coastal state where the private property rights extended to the low tidemark, there would be a greater likelihood of a takings claim because the background principles of property law of that state hold the intertidal zone to be private property. However, if a government were regulating the same intertidal zone in a coastal state where private property rights ended at the high tidemark, then there would be no deprivation of private property rights because the background principles of property law of that state establish public ownership of the intertidal zone.

The background principles of property law that exist in the coastal state help answer the question of whether a regulation is "reasonable" in relation to the police powers of the state; if the regulation interferes with historical private property rights, then there is a greater chance that the regulation is unreasonable; if the regulation does not interfere with historical private property rights, then the regulation may be reasonable. This is particularly true when trying to determine if a regulation is "reasonable" through a background principles approach rather than determining reasonableness through a different standard. However, the analysis of background principles of law does not necessarily dictate the ultimate conclusion of reasonableness as other factors, including the underlying existence of property rights within or outside social constructs, help establish the reasonableness of a particular act.* The combination of both background principles and property rights evolution are often found in judicial opinions and, although there is no uniform application of these standards, finding their existence in opinions can help us understand when they are being used and how they are being applied in a particular context. A summary of instances in which recent judicial decisions have described property rights

* The "background principles" of property law concept is a recurring theme in recent judicial opinions interpreting takings cases. However, the concept itself is somewhat problematic when viewed in light of "public nuisance" defenses to takings claims (a common law defense that often is applied to justify the reasonableness of the public regulation). The reason it is problematic is because it suggests, at least through certain interpretations of the doctrine, that certain property rights are defined historically in a way that does not change over time, even when new circumstances present a *condition* that is historical in nature. For example, sea level rise may be argued to be a new phenomenon that triggers the historical legal *principle* of nuisance; regulations aimed at protecting against sea level rise defend against a public nuisance, and protecting against public nuisances is very much a "reasonable" function of government under its police power. Because sea level rise is a newer condition by which a nuisance may be created, it can never be said that states have a historical principle of nuisance that is defined by sea level rise, and thus this represents a background principle of property law that justifies the states' action. Because sea level rise (as is occurring today) could not historically occur, it is impossible for the event to be considered a background principle of property law. However, nuisance itself is such a background principle. If the nuisance principle of sea level rise is accepted as the argument for justifying the reasonableness of the state action, then this is more likely a reflection of finding property rights existing within social constructs. To the contrary, if sea level rise cannot justify the basis for a public nuisance because it did not exist historically, then this ideological distinction seems to be supportive of the idea that property rights exist outside social constructs.

evolution under a takings analysis is presented now to inform the more directed analysis of takings jurisprudence in the following section. The principles discussed are placed into the context of coastal planning so that the relevance of the legal framework in relation to sea level rise adaptation strategy is made obvious.

Recall that the purpose of reviewing a court's rationale for supporting or invalidating government action is to understand when the regulatory police power of the state goes too far into private property rights such that the regulation is deemed unreasonable and thus a taking of private property rights. By following the rationale offered by the court, factors that influence its finding are identified and can help in understanding what kinds of regulations are more closely related to valid expressions of police powers rather than invalid takings of private property rights. As mentioned above, an initial focus of a court's review of state regulatory conduct is to analyze the actions itself in some detail. The goal of the court in focusing on the state action is to determine if the action (1) has a public purpose, and (2) whether the public purpose, if found, is based on a background principle of state property law. Where a background state property law is found, the court then seeks to see how the property law is being applied to the current situation. Where the property law is being applied in a manner that is in accord with prior uses of that law, then the law is generally found to conform to a valid police power. However, there is an additional analysis that occurs where the application of the valid police power, once confirmed, is then reviewed to determine its *impact* on property rights. Where the impact on the property right is found to be substantial, an additional set of inquiries tends to take place. First, if the property right being impacted is the same kind that has been subject to regulation under background principles of property law, then there is a greater likelihood that the regulation will be found valid. Second, if the property right being impacted is of a different kind of background principles of property law historically, then the impact is evaluated in relation to the effect of the regulation on the property right being impinged. Particular legal tests have been developed to determine when the impingement on property rights may be considered a taking at this point. Some of the tests include an analysis of the economic value of the property after the regulation. Other tests focus on the economic expectations of the landowner whose property is subject to the regulation. Still others balance the economic impact of the regulation against the public benefit, looking to factors such as the permanency of the regulation to evaluate the overall burden of the regulation on private property rights.

Coastal planning that involves adaptation strategies to sea level rise includes the kinds of government actions that run the range from very little impact on private property rights (mostly armoring, beach nourishment, and other techniques employed to protect against the incoming tide) to actions that have a significant impact on private property rights (buffer zones, no building zones, and other techniques that are more geared toward planned retreat strategies). Those strategies that impact private property rights are obviously the kinds that can lead to takings challenges. When considering adaptation strategies, coastal managers must have an awareness of the kinds of property laws that exist at the state level and attempt to link, to the extent practicable, the adaptation technique that may impact private property rights to a background principle of state property law. For example, the U.S. Supreme Court has recently interpreted background principles of state property law in relation to a state action of beach nourishment in *Stop the Beach Renourishment Inc. v. Florida Department of Environmental Protection*, 130 S.Ct. 2592 (2010). In that case, a government nourishment project that included adding sand between private property and the shoreline was challenged because the effect of the government project was to establish public ownership in the nourished strip of land, meaning that the private landowner's property no longer "touched" the shoreline.* In trying to determine whether a taking of property rights had occurred, the U.S. Supreme Court reviewed background principles of Florida property law to determine if Florida had a historical right to claim ownership of newly created uplands through publicly financed nourishment projects. The conclusion was that Florida did indeed have ownership over the newly created sandy beach because (1) the Florida statute at issue allowed for Florida to obtain ownership rights over beach nourishment projects to the extent of the land mass added by the public project, (2) this additional land was considered *an avulsion under Florida property law*, and (3) Florida property law allows for avulsions to accrue to the state and not the private property owner. In essence, because background principles of Florida property law considered the addition of land a kind that accrues to the state under background principles of property law, Florida was entitled to the ownership rights of the land.

The ultimate holding in *Stop the Beach* is not significantly important in itself for purposes of the points being made in this chapter, but rather the rationale employed by the U.S. Supreme Court Justices in coming to their decision is troubling from a policy standpoint for a variety of reasons. First, the reasoning of the plurality of justices takes a highly conservative stance on the inalienability of certain private property rights. This conservative stance retraces the foundations of property law to an existence outside the social constructs of government, meaning that it advances an argument that private property rights exist as a pre-political right (Rose 1999; Ely 2008). Second, the decision suggests that judicial opinions of the highest state courts can no longer be relied upon to interpret the extent of state property law, including the ability to update state property law to reflect changes in community standards and norms. As discussed below, the impact of these two holdings is significant because it presumptively retards the ability of coastal managers to effectively plan for and implement adaptation policies toward climate change by instilling doubt into where the line between Tenth Amendment regulation and Fifth Amendment takings lies.

---

* The private landowners claimed that one of the property rights that accrue to private landowners who abut the shoreline (*littoral* landowners) is the right to touch the shore. Thus, any government action that removes this right impedes on a fundamental property right of the private landowner and thus results in a "taking" of this right.

To place the impact of the *Stop the Beach* decision into context, some analysis of the property law principles being advanced by the Court should be summarized. We can place the history of property law into two broad theories: (1) the *Lockean theory* (named after philosopher John Locke) and (2) the *modern regulatory theory.* The Lockean theory places property rights outside the social constructs of government and society, making them a natural right (Ward 2010). Meanwhile, the modern regulatory theory holds that property rights are created through the social construct of society and thus through the modern regulatory state (Ackerman 1978; Freyfogle 2007). Under a Lockean view of property rights, the natural rights imbued into real property cannot be altered by the state. Alternatively, the modern regulatory theory holds that property rights are continually reshaped and refined to suit the changing needs of society. The two views can be seen as juxtapositions to one another in terms of where property rights derive: Lockean theory noting that property rights exist outside human-made institutions, and modern regulatory theory identifying property rights existing wholly within human-made institutions.

The way in which property rights are fundamentally defined impacts the way in which those rights can be influenced through later government action. *Stop the Beach* provides an example of this proposition as the holding in the case clearly indicates it favors a Lockean view of property rights. Under such a view, state actions that impede fundamental property rights are within the purview of a takings claim. Only when the state action impacts a characteristic of property that is shown to be historically outside the rights of private landowners may a state engage in regulation without being subject to a takings claim.

The problem with supporting a Lockean theory that certain property rights exist outside the regulatory state is that such a theory places these rights in a kind of vacuum, making them incapable of adapting to advancements in human understanding and concepts of the world. Importantly, a reliance on such a "static" understanding of property rights is not supported in the actual legal history of property law in the United States. First, there are ample examples in the literature where property rights are found to depend on the exercise of socially created public power (Freyfogle 2006) because "there is no such thing as natural property[;]... it is entirely the work of law.... Property and law are born together, and die together. Before laws were made, there was no property; take away laws, and property ceases" (Bentham 1840, p. 139). U.S. Supreme Court opinions have regularly upheld the idea that property laws, being a social creation of the state, change as social norms and preferences of society change; examples include *Board of Regents v. Roth*, 408 U.S. 564, 577 (1972) and *Phillips v. Washington Legal Foundation*, 524 U.S. 156, 164 (1998). Indeed, scholars have identified examples of history where property law has been altered to accommodate societal changes such as the industrial revolution (Freyfogle 2007, p. 54–55) and the rise of environmental law as a field of new regulatory powers applied to existing property rights (Ackerman 1978, p. 4).

Beyond the lack of precedent to support fundamental property rights as a "natural" right that is sacrosanct to regulation, there are real practical problems for policy makers to develop and implement innovative ways of managing sea level rise adaptation, particularly when those management innovations include land-use restrictions. A recent State Supreme Court case interpreting parts of the rationale in *Stop the Beach*—in particular, the defense of fundamental property rights—exemplifies how these real problems develop when a court is forced to analytically retrofit a property rights theory that does not comport to the actions and history taken in that particular state jurisdiction. While troubling and difficult for the courts, the consequences are at least as difficult, and very likely more difficult, for planners who have relied on different ways of managing coastal land-use expectations between users of the resource.

In cases such as *Stop the Beach Renourishment*, coastal managers are provided with information that helps them understand how adaptation techniques might be applied in the face of takings challenges; in Florida, the government clearly has the right to nourish the tidemark in such a way that creates new land between the preexisting landmass and the tidemark. The new land created is, under Florida law, public property and no compensation is required to be paid to the private landowners whose properties no longer "touch" the shoreline. In other coastal states, a similar claim of ownership to the newly created landmass may not be supported by background principles of property law in that state. Thus, the ability to claim public ownership rights of that property may lead to a legitimate takings claim for the private landowner.* Of course, in states where a background principle of law does not allow for *ownership* to accrue to the state, thus increasing a takings claim, the ability to engage in the nourishment project is still likely well-established under the police powers of the state. The main difference is the state may do the project but may not expect to gain ownership rights over the project.

Linking background principles of law within a state to coastal management planning is a fundamental way to analyze the likelihood of a takings claim being

* There is little doubt that the nourishment project itself is a valid exercise of the government's police power to protect the safety and welfare of its citizenry. This is particularly true where the nourishment project is being done to mitigate current rates of erosion and otherwise protect human development landward of the project. Indeed, private property owners whose properties touch the coastline directly benefit from the nourishment project because the sea is moved further away from their properties by the nourishment project, and thus the resulting dangers that accompany the sea are also further away. The issue here comes down to the property right of these coastal owners to "touch" the water. If this right exists under state property law, then a basis needs to exist in the state's property law that allows the government to infringe upon this right. So engaging in the nourishment project may be a *reasonable* articulation of the state's police power, demanding the resulting additional property become public property may be deemed *unreasonable* (too far outside the scope of the government's police power when balanced against the private property right) in a state that, unlike Florida, does not have a history of property law that allows such additions to accrue to the state as owner of the property. The question of whether the state can engage in this kind of action (filling its publicly owned submerged land) as a *property owner* rather than as a *regulator* to avoid a takings challenge is further discussed later in this chapter.

made and also the likelihood that such a claim may actually lead to a judicial finding of unconstitutional takings. Again, compensation can always be paid to sidestep the takings issue. Also, coastal planners can steer clear of regulatory actions that impact private property rights, thus avoiding the impact on property rights that triggers a takings challenge. However, considering many of the proactive policies for adaptation that include some consideration of retreat in the future (often the more efficient form of policy direction as described in Chapter 3) impact private property rights, it is likely reasonable and anticipatory that policy development will include government actions that inhibit private property rights at some point in time. It is this reality that requires a close inspection of takings cases. Understanding that takings claims involve a dynamic between the foundation for public actions, the public actions relationship to background principles of state property law, and the impact these acts have on private property rights, we now move on to an analysis of regulatory takings claims, which represent the kinds of issues presented when government policies impact private property rights.

*4.4.2 Regulatory Takings Analysis**

This section explores how federal constitutional limitations impact government's policy responses to sea level rise. To begin, we provide a background discussion of the Fifth and Fourteenth Amendments to the U.S. Constitution; specifically, the takings prohibition found in the Fifth Amendment and applied to individual state action through the Fourteenth Amendment is discussed, including an outline of the several kinds of government actions that can result in a taking of private property. This discussion is supplemented with relevant case law highlighting the major principles distinguishing government actions resulting in takings from actions that are insulated against takings claims. Specific emphasis is placed on how government conduct is categorized—for example, whether the government is acting as an owner of submerged land, a trustee of public rights in those submerged lands, or as a regulator of private landowners whose property abuts submerged land. The options between government taking analysis between a regulatory role and a nonregulatory role in advancing policy will be further discussed and elaborated on in the following section to this chapter. However, where applicable, the impact of the role that government takes in assessing the legal viability of a takings claim will be identified in order to note how a nonregulatory stance might mitigate takings claims in particular instances.

* A number of the points made in this section are incorporated from an article developed and published by this author in the *NOAA Sea Grant Law and Policy Journal* titled, "Climate Adaptation and the Fifth Amendment of the U.S. Constitution: A Regulatory Takings Analysis of Adaptation Strategies in Coastal Development with Application to Connecticut's Coastal Management" and published in Volume 5, No.1 pages 140–168 (2012), available at: http://nsglc.olemiss.edu/SGLPJ/vol5No1/McGuire.pdf.

4.4.2.1 History of Regulatory Takings

What is traditionally referred to as "takings law" deals with limitations on government actions that result in a deprivation of rights to private property.* The basis for this protection stems from the Fifth and Fourteenth Amendments to the U.S. Constitution.† Most states have adopted similar takings limitations in their respective state constitutions (Craig 2007; Craig 2010). It is important to note that not all government intrusions into private property rights result in a taking. Indeed, there are numerous examples where government actions affect private property rights but do not trigger constitutional scrutiny. As indicated earlier in this chapter, government regulations that are based on background principles of state property law and are reasonably related to the government's police powers are generally insulated from a takings finding. In addition, government can act outside its role as a regulator of private property rights, thus altering the dynamic between the public and private interaction such that a takings claim may be avoided. However, many government actions that are "regulatory" in nature and impact defined private property rights can be subject to takings claims. Thus, a review of takings case law is helpful in establishing when government regulation may move outside the comfort zone of traditional police powers and into the zone of unconstitutional takings claims.

More traditional takings claims tend to focus on *physical occupation*, where a government entity either condemns private land for a government purpose without compensation or where the government allows, through *regulation*, the permanent physical occupation of private property, again without compensation (Craig 2009, p. 149). The U.S. Supreme Court case that held that a government regulation resulting in the physical occupation of private property is a taking was *Loretto v. Teleprompter Manhattan CATV Corporation*, 458 U.S. 419 (1982). In the *Loretto* case, a city ordinance allowed for cable television boxes, owned by a third-party company, to be physically attached to privately owned residences, regardless of private landowner acquiescence. Because the cable boxes represented the placement of a physical structure by government regulation, the act was seen as publicly sanctioning occupation of private property. When a regulation allows for the ongoing occupation of a private property, a taking of that portion of the property may be found. As noted in *Loretto*, the government can sustain a permanent physical occupation

* It is presumed that the government actions being discussed here do not qualify as *eminent domain* actions, meaning instances where a government is purposefully and intentionally attempting to "take" private property for a public purpose and willing to pay fair-market value for the private property in question. Rather, all government actions described herein relate to acts that *result* in an impingement of private property rights without the intention of government to exercise its eminent domain power.

† The Fifth Amendment prohibits *federal* government action that would result in a "taking" of private property rights. The Fourteenth Amendment makes the prohibition on government takings applicable to the states.

as long as it meets the requirements of eminent domain: The permanent physical occupation is for a "public purpose" and the government pays the landowner just compensation (fair-market value) for the occupation. Whether by directly taking the property or by allowing for the occupation of private property, the government act of *possessing* the property—in whole or in part—distinguishes these forms of taking from regulatory takings where government does not actually possess (nor does government *intend* to possess) the property interest of the private landowner.

A *regulatory taking*—taking through regulation rather than physical possession—is a less clear form of takings analysis, especially when it relates to government planning that impedes the use of private property. Judicial opinions that began laying the foundations for regulatory takings started in the early twentieth century, where decisions began to recognize a taking could occur by regulation, meaning that government actions which did not result in the physical occupation of private property but otherwise inhibited private property rights could result in the unconstitutional taking of that property. In supporting the principle of a regulatory taking, the U.S. Supreme Court in *Pennsylvania Coal v. Mahon*, 260 U.S. 393 (1922), identified that while the regulation of private property was within the normal police powers of government, if regulation goes too far, it will be recognized as a taking. Thus, the principle supporting regulatory takings was acknowledged and defined by the courts; what remained to be answered from this principle was understanding how it applied in the context of particular factual settings: that is, when precisely does a regulation go too far?

The U.S. Supreme Court has recognized three general categories of regulatory takings: (1) permanent physical occupations as noted in the *Loretto* case above, (2) regulations that deprive landowners of all economic use of their property, and (3) regulations that deprive landowners of some use or value of their land. The first category of regulatory takings described here—permanent physical occupation—is seldom the basis for seaside landowners complaining about government policies impacting their use of land. Rather, most regulatory takings claims made by these landowners stem from the second and third categories of regulatory takings claims: those categories claiming some or all of their private property rights in land have been proscribed through government regulation.

As noted previously, powers not delegated to the United States by the U.S. Constitution are reserved to the states through the Tenth Amendment to the U.S. Constitution. Utilizing a wide variety of these reserved powers, including the police power, state governments have often passed laws limiting the use of private property. However, the ability of a state or local government to defend against a regulatory takings claim based on these police powers has been limited by the U.S. Supreme Court. Two cases, *Penn Central* and *Lucas,* helped to identify the factual differences between a reasonable exercise of police power where compensation is not required, and an unreasonable exercise of that same police power that goes too far into infringing private property rights such that a taking has occurred. Recall Figure 4.5 in this chapter and its visual representation of a continuum: On one side sits government's legitimate police powers, while on the other side of the continuum sits government actions that clearly result in a taking of private property rights requiring a public purpose and just compensation. Somewhere between these two ends exists a mixture of government action that may, under the circumstances, reach a point where the action goes just too far into private property rights. There is no exact calculus available to determine when a government act might reach this point. However, judicial opinions like the ones being reviewed here are important aids in helping to understand the kinds of facts that are necessary to move a government regulation into the realm of unconstitutional takings.

Pennsylvania Central Transportation Company v. New York City, 438 U.S. 104 (1978), established a three-part test to evaluate when regulation that deprives a landowner of some use or value of their land may result in an unconstitutional taking of private property without just compensation. Under the *Penn Central* test that was established in this judicial opinion, courts evaluate the following factors: the economic impact of the regulation on the landowner, the extent to which the regulation interferes with legitimate investment-backed expectations, and the character of the government action. *Penn Central* acknowledged valid exercises of government regulation can impact the value of private property, and the fact that regulation impacts private property rights does not automatically mean a taking of those rights has occurred. To determine if the regulation has exacted a taking of private property rights, there must be a context-specific analysis that incorporates the three factors identified above. So, for example, a regulation that is loosely connected to a reasonable government police power but significantly interferes with legitimate investment-backed expectations while having a substantial economic impact on the landowner is more likely to be considered a taking than a regulation that has little economic impact on the private landowner and is supported by a strong police power such as protecting public safety. In summary, the impact of the regulation must be balanced against the evaluative criteria identified above. When the weight of the evidence tips the analysis of those criteria in favor of the private property owner, then a regulatory taking may be found. If not, then a taking is likely not found. The question left somewhat unresolved in *Penn Central* is what specific facts might tip the scales toward a regulatory takings rather than toward a constitutional exercise of police power. This was partially discussed later by the U.S. Supreme Court in the *Lucas* case.

In *Lucas v. South Carolina Coastal Council*, 505 U.S. 1003 (1992), the owner of a buildable piece of coastal property was prevented from developing the property based on a new regulation by the state agency responsible for protecting coastal resources. The reason the state agency created the regulation was in large part to ensure that barrier islands protecting the coastline would be maintained as erosion of these islands tended to increase with development. The private landowner, Lucas, challenged the new regulation prohibiting all future development partially on the ground that the regulation deprived him of all viable economic use of his property.* In its ruling on the case, the U.S. Supreme Court affirmed the rights of states to enact regulations through their police power under *Penn Central*, but limited this power factually by stating that a regulation that removes *all viable economic use of property* results in a regulatory taking, thus creating the second category of regulatory takings described above. The question that remained was whether the South Carolina Coastal Commission regulation deprived Lucas of all viable economic use of his property. The Court did not answer this question. Rather, it remanded the case back to the South Carolina courts to make a determination based on the newly created category of regulatory takings rule – a regulatory taking is shown where the regulation removes all viable economic use of the property. The South Carolina Supreme Court determined that Lucas bought the property with the intention to develop it at some point in the future, paid a price for the property that included development, and thus the removal of the development right did exact a taking of Lucas' property because it removed all viable economic use of the property.

Although *Lucas* held a regulation that deprives landowners of all viable economic use of their property might result in an unconstitutional taking, the Court identified a major exception in rendering the decision. Specifically, the Court noted the following: "[w]here the State seeks to sustain regulation that deprives land of all economically beneficial use, we think it may resist compensation *only if the logically antecedent inquiry into the nature of the owner's estate shows that the proscribed use interests were not part of his title to begin with*" (*Lucas v. South Carolina Coastal Council*, 505 U.S. 1003, 1027 (1992)). The Court here is focusing on principles of state property law, indicating a regulation that removes all economic use of the property will be a taking unless that regulation reflects background principles of the state's law of property and nuisance already placed upon land ownership. As the Court in *Lucas* further explained in outlining the "background principles of law" exception to a regulatory taking, "A law or decree with such an effect must, in other words, do no more than duplicate the result that could have been achieved in the courts—by adjacent landowners (or other uniquely affected persons) under the State's law of private nuisance, or by the State under its complementary power to abate nuisances that affect the public generally…" Thus, the Court "boxed" the rule of depriving a landowner of all viable economic use in a number of ways. First,

* Lucas owned a piece of undeveloped land on one of the barrier islands that was subject to the new regulation. Prior to the regulation, the land was capable of being developed. After the regulation was enacted, Lucas lost the right to build on the land. Thus, Lucas was claiming that the property right to build on the land was "taken" through the government regulation. Lucas was attempting to prove the taking by linking the economic impacts of the regulation as one of the standards developed under earlier case law. The essence of the argument was that when the right to develop is removed from the property, the property loses all essential economic value and thus the regulation exacts too high an economic impact on the private landowner regardless of the police power being advanced by the government as the purpose of the regulation (in this case to protect citizens from the impacts of a coastal storm).

the Court clearly indicated that any regulation depriving a landowner of all viable economic use triggers takings scrutiny. However, that scrutiny must be balanced against a close examination of state property law. Where state property law principles clearly show the state is enforcing a right that it maintains, such as protecting against a public nuisance, then the economic impact on the landowner is mitigated by the fact that the land could never be put to the use contemplated as it would result in a public nuisance and private landowners would never receive a right to engage in such actions. Thus, the economic impact is put aside where the public action is clearly supported by background principles of property law. The South Carolina Supreme Court could find no background principles of public nuisance law to support the regulation on Lucas. Thus, a taking was ultimately found on his property and he was awarded fair-market value as just compensation for the impact of the regulation.*

To review, there are generally three categories of regulatory takings: (1) a *Loretto*-type regulation resulting in a physical occupation of private property, (2) a *Lucas*-type regulation diminishing all economic use of the private property, and (3) a *Penn Central*-type regulation diminishing some use or value of the private property. The *Loretto*-type category of regulatory takings is clear; if a regulation allows for the physical occupation of one's property, then a taking will result. In some ways, this can be seen as government taking affirmative steps to allow for the occupation of private property by some entity other than the owner. (In *Loretto,* it was the allowance of cable company equipment to be affixed to private property through regulation.)

Both the *Lucas*- and *Penn Central*-type regulatory takings are less clear; neither offer an obvious path regarding what kinds of police power actions might constitute a total deprivation of economic use nor, under the *Penn Central* analysis, do they clearly spell out how the interests of the respective parties are to be weighted.† It should be of little surprise then that governments should approach regulatory takings issues cautiously. What governments have seemed to focus on instead are the background principles of state property law as a means of defending their actions

* In other words, the regulation was found to be valid under the police powers of the state such that it could be supported through the eminent domain powers of the state, which require a public purpose and just compensation paid to the private landowner. However, there was insufficient evidence to show that the regulation was supported by background principles of property law (in this case nuisance) such that the state could avoid having to pay for the impact the regulation had on Lucas' private property rights.

† In the *Penn Central* opinion, Justice Brennan, writing for the majority opinion, noted that instead of setting forth clear standards, the Court had adopted "ad hoc factual inquiries" to determine whether a regulatory taking had occurred. The rule established in *Penn Central* is that regulation that passes some degree of damage to the economic value of the property will result in an unconstitutional taking, but there is no uniform measure offered of when this will occur. *Lucas* aids in the analysis by stating the obvious: Removal of all viable economic use of the property through regulation equates to a taking. However, defining precisely when "all viable economic use" has been removed in a contextual setting has proved difficult.

against regulatory takings claims (Byrne 2010). As identified in the *Lucas* case and pointed out earlier in this chapter, these background principles of property law generally include common law traditions adopted by the state, such as nuisance and custom. However, there are other general common law traditions and principles in property law that specifically apply to coastal issues including the state as owner of submerged lands and also as trustee of public rights within and around the coastal zone. By looking at these common law traditions and property right principles that apply particularly to coastal adaptation strategies, one can gain greater insights into the kinds of actions that may mitigate regulatory takings claims in the coastal zone.

4.4.2.2 Regulatory Takings and Coastal Adaptation

As the case summaries provided above make clear, regulating in the coastal zone is fraught with potential dangers surrounding the takings prohibition of the Fifth Amendment to the U.S. Constitution. In summarizing the different categories of takings identified above, some connections can be made between coastal adaptation techniques and case law interpreting those techniques in light of takings challenges. The purpose of this section is to provide a summary of how the case law principles described above might apply to different kinds of adaptation approaches, linking the general approach and specific techniques employed to the categories of takings described above where appropriate.

Separating adaptation approaches into the two main categories described in Chapter 3—staying and retreating—we can begin with a general approach that employs staying at the coastal region. Recall that the kinds of techniques employed in staying include protecting against the incoming sea and adapting to the incoming sea. When protecting against the incoming sea, there are three categories of techniques available: (1) fixed-position armoring, (2) flood protection, and (3) hybrid techniques. Collectively, the idea behind a stay-and-protect approach is to engage in geo-engineering projects that prevent or mitigate the effects of the incoming tide *without* impacting private land use to the degree practicable.

Generally speaking, the techniques involved with a stay approach fall more clearly into the zone of police powers that tend not to infringe upon private property rights to the degree that a takings finding is triggered as described above in the case law tests that have been established. This is because most of the techniques derive from affirmative government actions that do not ban certain uses of private property. For example, armoring techniques provide a structure that sits between the land and sea to protect landward areas. If the armoring is part of a publicly financed project, then the government usually uses its eminent domain powers to "take" private property where the structure is to be built, thus identifying a clear public purpose and paying just compensation to the aggrieved landowner within the guidelines established under the takings doctrine (McGuire and Hill 2012, p. 157–163). Beyond this overt government act of acquiring private land, where necessary, to build the seawall or other structure, there is little to no impact of the project on private property rights. In fact, construction of the armoring device itself is a signal by government that it is willing to engage in specific actions to protect private property rights, including limiting the extent of the landward approach of the sea to the point of the armoring.*

Other methods of protecting against the sea outside of armoring also limit the degree to which government directly interferes with existing private property rights. Breakwaters, jetties, and similar structures are all constructed in submerged land, which is publicly owned. The development of these projects can have some impact on private lands, for example by altering the natural flow of sediment down-current and thus impacting the aggregation of these sediments on certain private beachfronts. However, outside of this indirect impact, there is little to no direct connection between the use of these techniques and unreasonable interferences in private property rights, at least not to the degree that case law principles discussed above might identify the actions as "regulations" tending to remove private property rights. Again, this is primarily due to the fact that these kinds of actions are proactive government conduct that does not result in the banning of private use of property. Nourishment projects generally also fail to impact private property rights, as the government is proactively engaging in activities that enhance the uplands and protect existing private property. The only difference is when, as in *Stop the Beach*, the government is claiming ownership rights over the uplands created by the nourishment project and this claim results in separating private property rights to "touch the sea." When such claims are made, there is an impact on private property rights; landowners owned land that touched the water, but the nourishment project is removing that right by placing the new land under public ownership. Whether this act of depriving private landowners of a claimed property right is constitutional will depend on the background principles of property law of the jurisdiction in which the action is taking place. *Stop the Beach* informs us that in Florida there is no such private right at stake under the circumstances, and thus the state can claim ownership of this new strip of land without triggering a taking of private property rights.

Methods of retreat tend to follow a more *prohibitive* path in the policy techniques employed to manage coastal land use expectations. In many cases, the use of retreat techniques will mean changing the way in which private property owners have interacted with their coastal properties in the past. This change in behavior almost always includes government altering "rules," either today or tomorrow, in what government is willing to do in order to protect private landowner interests. There are a couple of ways in which government can go about altering these "rules"

* Recall that government is the owner of submerged lands. Further, government becomes the owner of newly submerged land that is the result of natural forces that slowly move the seas landward. Thus, by armoring, government is limiting its rights of ownership to future additions of submerged land in order to protect the private property rights of landowners upland from the armoring technique.

that draw important associations with takings jurisprudence. Government can certainly enact banning-type regulations that limit what people can do on their property. This can include using policy techniques such as buffer zones and setback requirements that are more stringent than those existing in the past. The effect of these kinds of rules will be to, often, make sections of private property undevelopable where they might have been developable under preexisting zoning rules. This is a similar kind of action as was engaged in by South Carolina's Coastal Commission in the *Lucas* case and thus brings up the possibility of a regulatory takings claim. Whether these kinds of actions will be considered takings depends on the ultimate economic impact the regulation has on a property. If it does not deprive the property owner of all viable economic use, but rather just some of the use, then it may not be seen as a taking of private property rights. In addition, if the actions of the government are supported by background principles of state property law, then there is a greater chance that the action will withstand a takings challenge.

Beyond methods of retreat that focus on banning-type regulations, there are more forward-looking regulations that may achieve the same goals of retreat but have little to no immediate impact on current land uses, while having a potential impact on those land uses over time as conditions change. The example of an overlay district was given as a comprehensive tool that allows changes in expectations on using land to be made today, but the actual expression of those changes does not take place until some future point in time. The difference here is that existing land uses are allowed to continue intact. Over time, as conditions change, the increasing risks of those conditions become borne by the private landowner rather than being insured by the government. For example, an overlay district may include the technique of a rolling easement aligned with a prohibition against armoring.* If sea level rise does not occur, then the rolling easement never takes effect. If sea level rise does occur, then the risks associated with that rise begin to shift from the public to the private landowner, depending on the extent of sea level rise. Because property rights will only be affected upon the condition of sea level rise, there is less chance that the enactment of the rolling easement can be argued as a taking of private property rights simply because existing private property rights have not been impacted by the future-looking regulation.

Whether retreat techniques include the immediate or forward-looking regulations, there is little doubt that retreat methods will tend to trigger takings claims more than government policies geared toward staying and protecting the shoreline. That said, it must also be acknowledged that retreat methods often provide superior means of managing expectations along coastal areas in the face of sea level

* The prohibition against armoring, including private armoring, may have to be established under background principles of property law. There may be some argument in certain coastal states that the right to armor against the incoming tide is a private property right, and any regulation that prevents armoring removes that private property right and thus constitutes a taking.

rise, particularly when management goals include distributing the greatest amount of "net benefits" to society as discussed and analyzed in Chapter 3. So there is a tradeoff between adaptation policy approaches where government policies geared toward staying at the shoreline may include techniques more insulated from takings claims, but offer sometimes perverse incentives in the face of sea level rise that actually increase risks and costs associated with remaining along vulnerable coastal regions. Planned retreat methods offer a means of altering land-use expectations over time that allow for the internalization of increasing risks by participants in the coastal zone in a way that may trigger takings challenges. Even so, such challenges may be worth any additional costs where takings are found or in avoiding takings through direct eminent domain procedures because they offer long-term benefits that may legitimately exceed these additional costs if and when they are incurred. Where the calculus is unclear, governments can often gain a clearer perspective of a potential takings challenge and its likelihood of success by understanding the evolution of judicially developed rules used to interpret the relationship between government regulation rights and private ownership rights in land.

4.5 Government Options

Government has a number of roles it can call upon to aid it in achieving policy goals. One role of government is that of regulator, where its police powers are used to manage expectations and behaviors of its citizenry, as well as to distribute benefits to those citizens. In its role as a manager of coastal areas, state government often seeks to manage expectations through its regulatory role, and this is especially true when looking at land-use management. A number of points have already been made on the importance of understanding government's role in relation to managing the expectations of citizens in coastal areas. The purpose of this section is to highlight the main differences between government's *regulatory* powers and other *nonregulatory* powers that it maintains in coastal areas, particularly its role as both an owner and trustee of public property including the rights and obligations associated with public ownership. The goal here is to give greater insight into the impact that different kinds of roles taken by government can have on coastal management practices in general, and sea level rise adaptation policy development and implementation in particular. Sometimes government can achieve a policy goal by utilizing a nonregulatory role in conjunction with other regulatory methods of land-use control. By doing so, government can mitigate the potential for a takings claim by, for example, engaging in a policy of retreat that combines regulatory techniques with nonregulatory actions to support the management of coastal land-use expectations. Thus, this section is offered to provide conceptual tools to the legal frameworks analysis described above in the hopes that coastal managers can explore these alternative roles as a means of achieving sound adaptation goals.

4.5.1 Regulatory

Viewing government as a regulator of private property suggests it is taking actions that fit within the definition of regulation. This means government is acting within its traditional role of legislation and administration. Government action identified within a traditional regulatory role tends to trigger regulatory takings considerations. To quickly review, regulatory takings can happen by physical occupation, removal of all economic value, or removal of some value or use of the private property. If we look at a continuum where the Fifth Amendment taking is placed at one end of the spectrum, and the Tenth Amendment police power is placed at the other end, we can begin to think about the kinds of regulations that might be more likely to result in regulatory takings (i.e., those regulations nearer the Fifth Amendment end), as opposed to the kinds of regulations that might reflect background principles of law well within the traditional police powers of the Tenth Amendment. The key here is to determine what kinds of government regulatory stances are most insulated from takings challenges.

As stated earlier, the Tenth Amendment reserves to state governments the power to enact legislation and regulations for the health, safety, and welfare of the citizenry. As also stated earlier, the U.S. Supreme Court has indicated there are limits to government's ability to claim a Tenth Amendment privilege; specifically, such a privilege cannot be claimed for regulations that result in a prohibition of all economically beneficial use of land. So where is the safe ground? It seems the safest ground is when a regulation, at its inception, is derived wholly as a rendition of preexisting state property rights. Or, in other words, the safe ground exists more clearly when government is enforcing background principles of state property law through legislation or regulation. A clear example of this is when a state government chooses to incorporate a common law principle into its statutory scheme, such as when a common law public nuisance doctrine is codified by the legislature. Enforcement of this codified public nuisance would generally not result in a regulatory takings claim because the action is based on background principles of property law, in this case the doctrine of public nuisance (Caldwell and Segall 2007, pp. 557–558). As long as the public nuisance is proven to have existed within the state's tradition of property law throughout history, the private landowner cannot claim a regulatory taking because they never had the property right to commit the public nuisance in the first place.

While the codification of common law public nuisance traditions may be a clear example, other examples of government acting in its role as regulator are less clear under a takings analysis. This is especially true when legislative and administrative policy deviates from common law traditions. Professor J. Peter Byrne sets forth a strong summary of how this occurs, especially in relation to managing policies aimed at dealing with property rights within the coastal zone under conditions of sea level rise. As Professor Byrne notes, "… nuisance litigation notoriously fails to adequately weigh the broad public interests present in environmental disputes"

(Byrne 2010, p. 634). Professor Byrne supports this proposition by making the case that current statutory and regulatory enactments related to coastal public policy reflect a greater understanding of the threats posed to the public due to climate change. This includes not only threats to littoral landowners themselves, but also to adjacent landowners when one engages in practices such as armoring to protect against rising seas. He also highlights the threats such actions pose on government as both a property owner of submerged lands and as a caretaker of environmental concerns including ecosystem services (Byrne 2010, p. 637–639).

What Professor Byrne is highlighting in his essay is the idea that statutory frameworks, including administrative implementation of statutory goals, is often a superior means of inculcating the advancements of society into a legal framework. In the context of sea level rise, advancements in science have allowed us to better forecast the likelihood and impacts of climate change (Intergovernmental Panel on Climate Change 2007). This kind of information would not be available under traditional common law. Thus, a modern government has tools of forecasting available to create proactive policies to protect the fundamental health, safety, and welfare of its citizenry, which is the hallmark of the Tenth Amendment to the U.S. Constitution. By focusing narrowly on common law doctrines as the basis of background principles, judicial review can place government in the awkward position of "retrofitting" forward-looking policies into a background principles framework of common law tradition.

Some court cases, where the government is prohibiting the use of armoring devices, seem to highlight this problem of relying on common law principles to justify forward-looking government regulations: for example, *Shell Island Homeowners Association v. Tomlinson*, 517 S.E.2d 401 (N.C. Ct. App. 1999); *Grundy v. Thurston County*, 117 P.3d 1089 (Wash. 2005). The courts have either chosen not to recognize a littoral landowner's common law right to protect their property from erosion or otherwise found the "common enemy rule" (a common law right) did not apply to sea water and therefore the private landowner had no property right being impinged by the prohibition against armoring. In both cases, the court felt compelled to analyze the regulatory takings claims through common law doctrine, and, in order to uphold the government regulation, the courts chose to find the nonexistence of a property right that prior precedent had suggested always existed (Byrne 2010).

In reviewing the courts analyses in these cases, some interesting points arise in relation to policy making from the government perspective. If government takes on the role of regulator, it presumably wishes to incorporate the most complete knowledge about dangers to the public when creating regulations. To mitigate risks associated with sea level rise, governments are looking to policies that place restrictions on coastal property rights, including the right to armor. The government may choose these restrictions to protect both the public as well as ecological values associated with natural coastal landscapes. However, when government acts proactively through proscriptive powers, it steps into the sphere of regulatory takings.

Once inside, government must rely on the courts' ability and willingness to update traditional views of background principles of property law (such as public nuisance) to include more forward-looking risks to the public, such as climate change. Recent case law suggests that courts are having some difficulty doing this, especially in jurisdictions where there has been little expansion of traditional public nuisance principles adopting forward-looking factors and environmental risks in the definition of what constitutes a nuisance.

The uncertainty created when regulatory tools for planning and implementing policy directions are threatened by the uncertainty of whether the regulation will be found to be an unconstitutional taking of private land. Governments unwilling to be dragged into elongated legal battles and potentially costly regulatory takings judgments may become recalcitrant in using regulation—their broadest policy tool in land-use planning—as a means to achieve important policy objectives. If this trend is proven, then even when putting aside other influences that might stay the government's hand in regulating coastal land use, there is a greater potential that innovative policy directions may be stymied or lost in responding to triggers of climate change, including sea level rise. There is some strong evidence in the literature that regulatory takings have indeed had a cooling effect on government's willingness to use regulation as a means of protecting against sea level rise (Lazarus 1993, 1426; Fisher 1993; Davis 1993; Lazarus 2008).

As stated above, regulation is a key policy instrument for coastal planners in being able to place limitations on how coastal property is used, and this kind of limitation is an important part of developing policy directions that respond to the threats posed by sea level rise (Repetto 2006). An inability to understand the confines of regulatory takings prevents coastal managers from using all the policy tools at their disposal. A failure to use all available tools limits the capacity of policy to respond to the dynamic threats of climate change, thus removing the ability for policy makers to act proactively in many cases, which means the best kinds of policy development may be resigned to reactive implementation after a disaster occurs. Recent court opinions delving into takings analysis provide a potential to reinforce uncertainty in coastal land management, which does not bode well for those hoping for a clearer sense of what actions coastal managers might take to avoid takings claims.

A recent Texas Supreme Court opinion, *Severance v. Patterson* (No. 09-0387), ___ S.W.3d ___ (Tex. 2012), decided after *Stop the Beach,* dealt with an analysis of private landowner property right expectations in the coastal zone. Because *Severance* is a state court opinion—an "inferior" court to the United States Supreme Court—the federal constitutional law issues related to property rights settled in *Stop the Beach* became binding precedent on the *Severance* court when rehearing the questions presented before it in early 2012. Most specifically for our purposes here was the U.S. Supreme Court's reliance on a Lockean theory of property rights in *Stop the Beach* to support its holding that fundamental property rights exist and cannot be undone later by changes in societal expectations; rather, only historically provable background principles of property law may establish the basis for justifying government regulation that interferes with historically proven private landowner rights. An overview of the *Severance* case is provided below and then the majority opinion is analyzed in light of the precedential framework regarding property rights developed under the U.S. Supreme Court opinion in *Stop the Beach.*

Severance is a case about how an existing state law and state constitutional amendment protect the right of the public to access the nearshore, and how that right of public access is impacted by different legal interpretations of property rights. The competing property rights at issue in this case are the aforementioned right of the public to access dry land that lies immediately adjacent to the water, and the corresponding property rights of private landowners, including a right to expect that current established property rights will not be impeded by future changes in circumstances. The private landowner property right at stake here is a bit difficult to conceptualize but is broadly based on what constitutes reasonable property right expectations in a dynamic setting.

To place the property rights at issue in *Severance* into context, we must begin by understanding traditional notions of property rights *between* respective property owners and how those rights create expectations that are impacted by changing circumstances. We can begin by observing that property lines are generally assumed to be static, or not capable of changing over time; the precise area of land that one owns is calculated and defined with geospatial precision, more so today than in times past (Killenbeck 2002). Putting certain exceptions aside, the general rule is that a private person takes title to land knowing its precise dimensions and also knowing that, subject to certain Tenth Amendment police powers and any express restrictions placed into the title prior to the landowner receiving title (rights granted to others through easements, covenants that restrict how the land can be used, etc.), those dimensions and rights cannot be changed, unless changes are voluntarily undertaken by the landowner or done in accordance with proper eminent domain powers of government (Killenbeck 2002).

The question presented in *Severance* is whether a private landowner's property can be subject to future restriction when dynamic and potentially unforeseen events such as sea level rise cause public rights to "move" onto the preexisting private property of the landowner. Based on what has been stated above under precedent such as *Lucas,* the answer to this question is a qualified yes under the following condition: as long as the "future restriction" is based on a background principle of property law (recall that background principles of property law prevent the owner from gaining certain rights to the property, such as the right to set fire to the property, endangering neighbors and their lands—a "public nuisance"). Because private landowners never receive the right to engage in a public nuisance on their land, new or prospective government actions—particularly regulations—that restrict the private landowners' use of their property cannot impact the ownership rights of private landowners because private landowners never obtained the right to commit a public nuisance when they received title to their land. However, the answer here is somewhat dependent on what is truly meant by the term "background principle of law," and whether *new* circumstances can provide an equivalent basis for meeting the background principle exception enounced under *Lucas* and reemphasized in *Stop the Beach*.

Severance contained a majority opinion and several dissenting opinions. This is not uncommon in courts of appeal where individual justices might agree on the complete reasoning of an opinion, they might agree on the conclusion stated in the opinion but have different reasoning (a "concurrence") to support the conclusion, or they might disagree with the conclusion and some or all of the reasoning, in which case they usually file a dissenting opinion. Focusing on our question of whether sea level rise can be labeled a "background principle of law," the majority opinion in *Severance* concluded with an "it depends" finding. The majority were clear that sea level rise changes ownership of the land, shifting title of the newly submerged land from the private landowner, if any, to the public. However, the other property right at stake under *Severance*—the right of the public to access the nearshore—was left unresolved. The majority opinion essentially bifurcated the concept of background principle based on the *rate* at which sea level rises, and it appears that a close inspection of the rationale supporting the majority opinion makes this bifurcation along the same reasoning displayed in *Stop the Beach* regarding assumptions about underlying property right principles. The connection between the majority opinion and this underlying property right assumption is drawn out in greater detail below.

Regarding the public's right to access the sea in Texas—codified in both the state's constitution and statutory decree—the *Severance* majority distinguished expectations between private and public landowners at the sea's edge based on the historical principles of *avulsion* and *erosion* as they relate to changes in boundaries between water and land. In summary, the *Severance* majority held that a slow natural progression of the sea landward through erosion would allow for the right of public access to move with the sea without impeding private landowner rights. However, an avulsive event—a sudden and substantial change in the boundaries between public and private land—does not automatically allow for the corresponding public right of access to "roll" with the event. In this case, the right of public access must be proven to exist adjacent to the new water line boundary under some other property right precedent beyond an erosion-type finding; the right of access is not presumed to "roll" with the change in boundaries when the sea level boundary changes suddenly and significantly in scale.

The relevance of the majority holding in *Severance*, a decision that inculcates the property rights principles supported in *Stop the Beach*, is that it expresses a trend in legal analysis that makes it difficult to reconcile the kinds of associations between property rights and societal expectations that are necessary when thinking about land-use policies grounded in the capacity to adapt to changing circumstances, particularly the kind of circumstances brought on by climate change. Sea level rise at the coastline is presently a fact for most low-lying coastal regions (U.S. Global Change Research Program 2009; Titus et al. 2009). The fact of sea level rise complicates traditional policy directions, particularly when ecosystem services and functions are connected to our thinking about how to respond to sea level rise. For example, as noted throughout this text, armoring to prevent the encroachment of the sea may lead to short-term protection of existing human structures, but it does so at the mid- and long-term expense of natural coastline habitats, stripping the coastline of the ability to maintain and develop natural coastal habitats as the sea rises for the indefinite future (Titus 1998; Titus et al. 2009).

Judicial opinions such as *Stop the Beach* and *Severance* defend a kind of property right principle that is juxtaposed to the reality that society experiences through climate change and its coastal impacts such as sea level rise. In essence, these cases support a property right principle where cost-shifting for the impacts of climate change is being made at the expense of the public for the benefit of private landowners who happen to own land along coastal areas. While government often chooses to provide cost-shifting measures to private citizens when disasters occur—for example, declaring a hurricane-ravaged area a disaster for purposes of public money to mitigate impacts; supporting a property right ideal that demands government compensation for acts of nature substantially limits the options available to policy makers when considering adaptation strategies along coastal regions. This is even more true when the rationale supporting a judicial decision does not comport with historical property right interpretations.

An example of this disconnect is found in the legal analysis supporting the majority decision in *Severance*. The public right of access to the water—a "public easement"—was not found to automatically inhere to the shoreline after an avulsive event. However, the legal history of easements as a property right clearly identifies the *purpose* of the easement as the main criteria in judging whether an easement continues to exist after some intervening event (Freyfogle 2007). Easements are non-possessory property interests that authorize the person benefited by the easement to use the property of another for a particular *purpose*. Access easements are a clear example of this "purpose" concept. The ability to access another person's property is always a means to some other end; access to the property is not the goal but is necessary to accomplish the goal—for example, accessing another's property in order to get to a third property that can only be reached through the easement. The majority in *Severance* place a *geographic* limitation on the easement even when the *purpose* of the easement still exists after an avulsive event; the movement of the sea suddenly inland still creates a sea–land boundary after the event ceases, and the *purpose* of accessing the shoreline—albeit a new shoreline—by the public still remains.

The *Severance* majority seems strongly influenced by the plurality holding in *Stop the Beach* in taking a Lockean view of property rights when determining the relative public-private interests in the coastal zone. The decision is supportive of private property rights irrespective of precedential support for the public's right of access—codified by statute—to the nearshore. It appears the court is hesitant to wholly submit private property rights to the purpose of a public easement as that easement "rolls" landward due to rising seas. The fact that the court must go to great lengths to differentiate its decision from prior precedent hints at a larger frustration between the realities of dynamic changing conditions under climate change, specifically sea level rise, and the need to preserve the ideal of private land as having certain definitive rights that are sacrosanct and protected against outside influences. The frustration centers on a change in expectations about background environmental conditions. Many of our expectations surrounding private property are bound to an underlying assumption of relative *stability*: stability in the belief that the law will protect the expectations of private landowner rights associated with real property, and stability in the geographic demarcation of our property interests. Both of these assumptions about stability are at issue when we overlay a dynamic environment that is capable of altering those assumptions in an instant.

Looking at this issue through a slightly different *lens*, we might say that property rights are really about creating assurances; most people would be less willing to invest in private property where there was less assurance as to the rights the person was getting in the bargain, mainly because the rights associated with private property help define the "value" of the property in an economic sense (Anderson and Hill 2003). Institutions, both public and private, have traditionally stood behind reasonable assumptions of stability in property rights, even in areas such as the coastal zone that are more dynamic and subject to change over time than inland areas (Fischel 2003). However, the amount of change that is assumed in these areas is moderate and calculated based on past experience: 50- to 100-year floodplains are calculated making assumptions about flood patterns from past trends (Schumann 2011); hurricane and tornado predictions for insurance purposes are also based on analyzing probabilities stemming from using the past as an indicator of the future (Jagger, Elsner, and Saunders 2008). If we could not establish a reasonable presumption of risk factors based on past experience, then questions surrounding the stability of certain property itself would be at issue, certainly property located in sensitive regions such as low-lying coastal areas.

Current judicial interpretations of legal frameworks affecting private property rights, particularly *Stop the Beach* and *Severance* described above, suggest that courts are moving in a direction that limits the ability of public land-use managers to develop policies that can adapt to sea level rise when those policies impact "fundamental" property rights and the policy is not supported by "background principles of property law" as outlined in the U.S. Supreme Court cases above. Any regulation enacted by government where that regulation is *not* determined to be supported by background principles of property law, but the regulation does result in the loss of property rights of the private landowner, is likely subject to a Fifth Amendment takings claim. While the result of such a challenge by private landowners cannot be predetermined with guaranteed precision, cases such as *Stop the Beach* create precedential interpretations that favor a takings finding without special exceptions noted. Because climate change can arguably create impacts of the kind not experienced historically—including increased rates of intensive storms resulting in avulsive events along the coastline—the background principles defense may provide even less hope in policy making after *Stop the Beach*. Public policy makers must look outside traditional notions of *regulation* to other ways in which government can advance its interests in the coastal zone for the benefit of the public at large. Government as the *owner* of public lands as well as the *trustee* of those public lands provides two examples of government roles outside the regulatory sphere of traditional government power. This chapter now looks at policy approaches that embrace these alternative government roles as ways of mitigating the impacts of the takings legal frameworks, including the recent court decisions interpreting takings claims.

4.5.2 Nonregulatory

The legal frameworks identified above center around certain government actions that are found to violate the Fifth Amendment of the United States' prohibition against the taking of private property without a public purpose and the payment of just compensation. At the heart of this idea of government conduct are actions taken by government that can be categorized as "regulatory" in nature, meaning that government is engaging in its police power functions to limit private property use (Leuchtenberg 2002). As a simple example, if the government is acting in its capacity as regulator, then it is more likely that regulatory takings are triggered because the government is clearly engaged in *regulation* (one of the criteria to meet a "regulatory" takings claim). However, government has capacities beyond that of a regulator; government also acts as the owner of property as well as having trustee obligations as relates to public property. These two alternative functions of government—property owner and trustee—are examined here in order to identify and explore alternative approaches to policy development that avoid the regulatory trigger for legal frameworks identified above, thereby removing difficulties for public managers in developing and implementing policy instruments responsive to sea level rise.

4.5.2.1 Government as Owner of Property

Similar to any private landowner, government maintains proprietary interests in the real property it owns (Sax 2010). As a landowner, government can advance its ownership interests, including defending its property rights against adjoining landowners. In addition, government can take proactive steps to protect its interests that attach to its ownership of land. Most importantly, none of these actions necessitates a "regulatory" stance from government, thus avoiding regulatory takings concerns.

The purpose of government is to advance the interests and well-being of its citizens. Government can effectively engage in this process through statutory frameworks, where societal needs are met in the development of regulations through statutory and administrative functions of government, often aided by advancements in science and otherwise an increased understanding of the causal relationships between our environment and human well-being (Fisher 2007). Climate change is an example of this process; government gains insights into how the environment is being impacted through research that better informs our understanding of causality, and these insights help form the foundation of government action through a regulatory framework of goals passed through legislation and then implemented through the administrative functions of government. However, this kind of activity is precisely the kind of "regulatory" stance that places government in the position of defending Fifth Amendment takings claims through regulation (Byrne 2010). In addition, the tenor of recent judicial decisions identified above shows little capacity for government to claim *new* threats as a basis for engaging in regulation that will escape a takings analysis under a "background principles of law" defense.

Government enforcing its rights as property owner removes the regulatory takings questions allowing government to adopt proactive policy directions that achieve similar goals: Protecting the welfare of the public through actions aimed at mitigating the impacts of sea level rise. The major difference is that government is choosing to "frame" its actions from the reference point of property owner rather than that of traditional regulator. In essence, government is attempting to engage in some of the same actions it might employ under its regulatory power; however, it is strategically choosing to employ these policy initiatives through power granted to it as landowner, rather than regulator, in order to avoid the regulatory takings claim altogether.

There are limitations as to what government can accomplish as a property owner instead of the broader scope of powers it might have at its disposal as a regulator in the public interest, and this is certainly true in the kinds of actions it might take to respond to sea level rise as a property owner. However, government can accomplish many of the same goals it might otherwise attempt to accomplish as a regulator of nearshore property. For example, consider that government wishes to prevent private littoral landowners from erecting seawalls and other barriers to prevent the rising sea from encroaching further inland. Under a "regulatory" stance, government might simply pass a regulation that bans the ability of anyone to build such barriers along the sea's edge. If private littoral landowners had been allowed to engage in such activities in the past, then it is very likely they will claim a regulatory taking of the private landowners' right to protect their property (including the future loss of property) through this regulation. While the ultimate success of such a claim depends on a variety of factors that are specific to the history of common law and judicial precedents within the jurisdiction, it is undoubted that the regulatory stance taken by the government places the controversy directly within a regulatory takings context.

One alternative path for government, rather than taking a regulatory stance to control barriers along the sea's edge, is to enforce its rights as the owner of submerged land. Assuming that rising seas cause the slow erosion of upland, and further assuming that once inundated the previous uplands now become submerged land, then that new submerged land is property owned by the public; by operation of law, the newly submerged land becomes public land. Government can take the role of landowner to enforce its rights to the newly created (or soon-to-be created) submerged land by seeking to enjoin any private landowner from creating barriers to hold back the sea, effectively engaging in an action that prevents the public land from moving landward. In such cases, the government has removed itself from the superior role of regulator-over-citizen to the equal role of landowner-landowner. As a holder of land with certain rights, government can seek judicial intervention of a private landowner's actions that might impact the rights of government as a landowner. The question now is no longer one of constitutional protection from government actions, but rather becomes one of two property owners seeking to have their respective rights and interests resolved. What is more enticing about this proposition is that courts are evolving to use a *reasonableness* standard to resolve disputes *between* landowners (Sax 2010). Such a standard looks at the competing interests of the property owners and seeks to redress the issue by balancing those interests. Under such an analysis, governments can bring into the discussion many of the issues raised by climate change, sea level rise, and the need to rationally resolve the interests that derive from the dangers imposed.

As noted above, the idea of government acting solely as a property owner to engage in policy for adapting to sea level rise has limitations. For example, certain jurisdictions where property rights are at issue follow a "common enemy doctrine," allowing owners to protect themselves against rising waters even at the detriment to an adjacent owner (e.g., supporting armoring). Other jurisdictions favor the "civil law rule," requiring the natural flow to occur unimpeded (supporting advancement of submerged land). In all the examples presented here, none offers an ideal solution *between* property owners. However, the key difference is that both parties are placed in a context where the focus is on resolving competing property rights, rather than the regulatory setting that creates positions between the public and private interests and invokes claims of regulatory takings.

The role of government as property owner is one way to move the focus away from regulation and toward creating equitable outcomes. While it is unlikely that a government can adopt holistic policies to deal with sea level rise based *solely* on its status as a property owner, allowing for the distinction to be part of the policymaking process can have significant advantages in how a government entity proceeds in implementing a proactive policy geared toward adapting to the effects of sea level rise brought on by climate change. We next look at the government's role as trustee of public property, where government is obligated to take actions to protect the public resource for the benefit of the public at large.

4.5.2.2 Government as Trustee of Public Rights

For over a century, *Illinois Central Railroad v. Illinois*, 146 U.S. 387 (1892) has established that states hold title to submerged lands within navigable waters, and the states hold these lands in trust for the benefit of the public. Public benefits generally include traditional rights of navigation, commerce, and fishing (Kalo et al. 2007). Importantly, the trustee obligations of the state in managing these public rights are not easily alienable. Thus, even when the state attempts to transfer the private interests (*jus privatum*) in submerged lands to a private party, the public rights of access and use remain, and the state's duties as trustee also remain. Collectively, this duty of the state to manage the public rights in navigable submerged lands is known as the *public trust doctrine.*

The public trust doctrine places affirmative obligations on government to actively protect and manage public rights in submerged lands. These trustee duties may include a duty to protect the nature of the public rights associated with ownership of submerged lands: Traditional public rights include fishing, bird hunting, and navigation. Certainly, these traditional public rights could form the basis upon which government could identify "background principles of law" in taking action to protect the public interest (Kalo et al. 2007). In states where public trust rights have been expanded within the coastal zone, it is reasonable to suggest that the trustee obligations of government to protect those rights expands accordingly (Craig 2007; Craig 2010).

Public trust doctrine as a "background principle" of state property law is an important factor for governments to consider when defending against regulatory takings claims. In essence, where the government is acting through its obligations as trustee to enforce public rights in the coastal zone, the actions related to that enforcement are grounded in background principles of state property law; therefore, the actions cannot be seen as triggering a regulatory taking because the state is simply protecting the public rights that flow from the doctrine itself. As some legal commentators have noted, "…the doctrine provides the most fundamental basis for responding to the threats of coastal armoring…" because the doctrine reflects the role government plays in ensuring established public rights in the coastal zone (Caldwell 2007, p. 552). This is an important conceptual difference in how the state approaches coastal issues—the difference between enforcing *regulations* and enforcing *trustee obligations.* When the state acts as a trustee, its actions begin with a "background principle" of property law; thus, all actions that follow can be linked back to the original trustee obligations. A state can argue that it is simply carrying out nondiscretionary trustee duties. Similar to the arguments above where the state adopts the role of "property owner" to defend its actions, the public trust doctrine creates an important conceptual distinction between the state choosing to regulate private land, which can create regulatory takings issues, and the state simply engaging in its obligations as trustee on behalf of the public.

The extent of trustee duties of government under the public trust doctrine, and thus the limitations of this government role, varies by state, depending primarily on the extent to which that state has developed its public trust doctrine interests (Craig 2007; Craig 2010). Some states have expanded their public trust rights to include access rights that "roll" with sea level rise, while others have included ecological values as important attributes of the public trust in coastal areas (Caldwell 2007). For example, if a government recognizes the rights of the public to access recreational opportunities, including traditional beach use of the nearshore, then it seems reasonable that the same government can take steps to protect the maintenance of the sandy beach. This may include state nourishment projects to mitigate the erosion of sandy beaches, as well as defending against private landowners attempting to armor when armoring is clearly shown to aid in the destruction of public rights. Again, the important factor here is linking the state action to accepted public rights in the coastal area. For some states, the public trust doctrine will serve as a more expansive tool against regulatory takings claims because of expanded public rights in the coastal zone. In other states, the use of the public trust doctrine may be more limited because of limitations in the development of public trust doctrine rights.

4.6 Conclusion

Analyzing policy directions through legal frameworks, particularly constitutional laws, provides insights into the capacity for coastal governments to adapt to sea level rise. This is particularly true when these constitutional law frameworks serve as the backdrop for analyzing property rights interpretations through judicial opinions. This chapter reviewed property rights issues along the coastal zone, and judicial interpretations of those property rights in light of Fifth Amendment takings prohibitions and Tenth Amendment police powers of coastal states, and hinted at some of the impacts that the property rights legal framework described can have on policy directions aimed at adapting to the challenges of sea level rise. A few potential implications of these impacts are now summarized.

First, it should be noted that legal frameworks are an important means of identifying policy-related issues (Rosenbloom, Kravchuk, and Clerkin 2008). By identifying the legal structure surrounding a policy issue, one can gain insights into the regulatory dynamics that can impact policy directions. In this chapter, constitutional law, common law of property, codified law, and judicial opinions (case law) were several legal frameworks identified to highlight the role that these frameworks can have on government's ability to manage sea level rise in the coastal zone by managing expectations of real property owners within this area. The Fifth Amendment to the U.S. Constitution was identified as an inhibitor to policy development where the proposed policy results in a "regulation" of private property rights such that a "taking" of private property rights is found to have occurred. Importantly, the taking of such rights can be found where the *effect* of the policy in practice results in a taking of private property rights even when there is no *intention* within the policy itself to actually "take" private land. The fact that well-intentioned policies can result in a government violation of constitutional protections is in itself an important consideration when thinking about the role of legal frameworks toward policy initiatives in coastal planning.

A second implication is the current state of case law developing around coastal zone property rights and the impact these judicial interpretations can have on the future of coastal zone policy development that infringes on private property rights. The history of *regulatory takings* jurisprudence has held that a government action may be deemed a taking even when that action does not intend a taking under the following circumstances: (1) where government action allows for permanent physical occupation of the private property, (2) where government action removes all viable economic use of the private property, and (3) where government action removes some of the economic value of the private property. One major exception to these regulatory takings cases occurs when government can prove that its action impacts the private property in a way that does not infringe upon an actual property right that is owned by the private landowner; in essence, a government action must impact an actual and realized property right of the private landowner in order for a regulatory takings claim to be made.* Because of this requirement of landowner rights, the question of how property rights are being defined in the coastal zone is important because this question helps policy makers understand how policy options might impact property rights and whether those impacts touch upon a right that can potentially bring about a regulatory takings claim, thus impacting the legal validity of certain policy choices.

The *Stop the Beach* opinion provides some compelling evidence that the U.S. Supreme Court is hesitant to take a view of property rights that is favorable to initiatives in the coastal zone allowing for proactive and expanded policies toward sea level rise when those policies impact private landowner expectations. *Stop the Beach*, and *Severance* as an application of some of the real property principles discussed in *Stop the Beach*, suggests that U.S. courts will tend to favor a property rights framework that identifies property rights as existing outside the social constructs in which they are implemented. If real property rights are pre-political in nature, then the court's reasoning is that regulation of these rights by government must be in accordance with certain "background principles" of property law allowing for the deprivation of these rights from time immemorial. Without such a historical basis for impacting private real property rights, government conduct that is regulatory in nature and impacts these pre-political property rights may constitute a taking in violation of federal (and possibly state) constitutional principles.

A troubling interpretation of this pre-political stance on property rights is that it requires historical referencing to account for the kinds of actions that may be considered "background" principles of property law. However the historical reference is not attached to the *categories* of actions impacting property rights, but rather the *kinds* of specific acts. For example, the U.S. Supreme Court held in *Stop the Beach* that Florida's historical practice of beach nourishment and subsequently claiming title to the newly created sandy beach area was *not* a taking because historically Florida had never ensured the right to "touch the water" (the property right claimed by the private littoral landowners never existed as a background principle of Florida property law). Had this right existed historically, it is likely that the Court would have found a "taking" of this right through Florida's actions, even though the *basis* for the action was to proactively prevent the potentially hazardous impacts of a rising sea and the encroaching storm impacts this would bring on the private littoral landowners.

* Also, implicitly, the government act must be framed as a "regulation" in order to trigger a *regulatory takings* claim.

In essence, this pre-political property right formulation limits the ability of government to claim a *public nuisance* background principle of law as the basis for engaging in actions today that would mitigate, if not prevent, reasonable forecasts of likely future public harm without having to pay private property owners for the privilege of protecting against such harms. In short, this pre-political property right formulation is turning sound government planning toward adaptive sea level rise strategies into de-facto eminent domain procedures, meaning that government must not only consider the costs of implementing adaptation strategies, such as the costs of a beach nourishment project, but it must also consider whether to claim up front an eminent domain procedure and pay for the private property right being "taken" in the process, yield the accrued benefits back to the private landowners, or otherwise plan for the costs of defending against takings claims and the possibility of paying for the value of the property right lost when a judicial determination of a taking is made. None of these options provides ideal incentives, or assurances, to government when considering planning for sea level rise adaptation strategies. Rather, the incentives displayed in these judicial opinions seem to suggest that government allow for sea level rise to happen and then simply respond to the disaster where the response can reasonably be couched in public health and safety terms and more directly avoid a takings claim.

A third implication to consider is how government can better respond to the judicial frameworks that seem to be evolving toward coastal property rights; in particular, does government have options for proactive strategies where there is a reasonable likelihood that these proactive strategies will not be subjected to takings claims as identified above? The answer offered in this chapter is a qualified *yes*, with the better chances of avoiding regulatory takings claims coming from government looking to its *role* when engaging in coastal policy planning, and then using its role to define how it goes about developing policy responses toward sea level rise. When regulating, government must pay close attention to the background principles of property law that are being impacted by the regulation. If there is clear evidence that government is simply reinforcing a right to protect against public nuisance in its regulatory stance, then there is a greater likelihood that its actions will not be deemed a taking. This is most clearly true where government can link the action taken to claimed "property rights" by private landowners that, under state property law principles, are not actual property rights owned by the landowners under the circumstance.* Otherwise, a regulation that is not clearly supported by background

* As an example, private property owners have a right to engage in a variety of activities on their land. One such right includes, under *allowable* circumstances, the right to create fire on the property (to burn leaves during prescribed seasons with necessary permits). This general right to create a fire on one's property is bounded however; one never has the right to create the kind of fire (in size and intensity) that is capable of and likely to destroy neighboring homes. Thus, the property right to create fires under specific circumstances *never* extends to a right to create a fire that will cause a public nuisance. The right to create a fire is bounded under public nuisance limitations. Linking coastal regulations to similar limitations creates an increased chance the regulation will be found to be outside regulatory takings challenges.

principles of law and shows deprivation of private property rights has a greater chance of being viewed as a taking of private property rights without compensation. Recent court cases dealing with regulations at the coastal zone support this interpretation and general view where, absent clear indicators of background principles of law, government regulations may have an uphill battle in being clearly seen outside a regulatory takings sphere.

As noted in the body of this chapter, government has roles outside that of "regulator" in the coastal zone; government is an *owner* of property in the coastal zone, and government also has *trustee* obligations to the public citizenry in managing its ownership rights in the coastal region. Government can take advantage of these other roles when considering policy development options. In particular, government can often avoid or mitigate the implications of takings jurisprudence by stepping outside its regulatory role. For example, government has the right to advance and protect its property rights as the owner of submerged land. When advancing these rights, government is not "regulating" the actions of private landowners but, rather, relying on established property rights principles. Advancing a policy directive from this approach can be advantageous in ways beyond that of avoiding takings claims. One advantage is that government as landowner can more easily negotiate rights with private landowners to come to a compromised solution that mitigates the impacts on both parties. Like abutting landowners deciding how to handle a dispute along a common property line (e.g., how to handle tree growth that impacts both properties), the opportunity to negotiate a mutual solution is bolstered by the fact that both parties sit as equals regarding property rights issues. If the issue cannot be resolved amicably between the landowners, judicial frameworks for reviewing these kinds of disputes *between* property owners often focuses on balancing the rights at stake and seeking a reasonable solution to the issue. In either case, government is placed in a superior position when the rationale for the policy direction is based in property ownership rights rather than on "regulating" in the public's interest.

Government's role as *trustee* has similar advantages over its role as a "regulator" when considering takings implications. As noted in the main body of this chapter, government has nondiscretionary duties that attach to its role as trustee of publicly owned lands, and it can rely on these duties when taking actions that might impact private property rights as a consequence of enforcing the public rights in the resource. For example, government may take actions today to protect well-established public trust rights based on state property law traditions; if and when those actions impinge on private property rights, government can defend the impact on the private landowner as a consequence of government protecting a public right rather than defending the impact from a regulatory stance that can trigger the takings/property rights issues described above. There is no definitive guarantee that government can always defend against a takings challenge by simply claiming that its action was spurred by trustee obligations. However, the takings challenge will have to overcome the basis for the government conduct—advancement of a well-established public property right—rather than being afforded a presumption that the government action is aimed at "regulating" the conduct of the private landowner. The fact that the government action is spurred through affirmative conduct based on an obligation under the law, rather than the action being seen as a restriction of private landowner rights, can have a significant impact on the overall validity of the government act, as well as the negotiation space between government and the private party when discussing the various rights involved in the dispute.

In conclusion, there are serious implications for coastal policy makers when the impact of proactive adaptation strategies toward coastal land use are viewed through the regulatory lens of government power as interpreted by recent judicial decisions by the U.S. Supreme Court and inferior courts interpreting the Supreme Court's precedent. A major cause of this problem seems to be based on a foundational premise being accepted by a majority of the bench, specifically that real property rights are pre-political in nature and thus exist outside the framework of societal expectations. The result of this foundational premise on real property rights is that coastal management policies deemed to be "regulating" private property rights as a means of achieving coastal sea level rise adaptation goals are subject to takings challenges under the Fifth Amendment to the U.S. Constitution. One way of mitigating this impact is for coastal managers to develop policies that derive from powers of government outside the role of "regulator." Two powers in particular related to coastal issues are the powers of government as the *owner* of submerged lands (property owner rights) as well as the powers of government as *trustee* of the public resource (trustee obligations). Together, these two alternative roles of government can serve as the basis for policy development in achieving the goals of sea level rise adaptation while mitigating the chance that these policies will be challenged as an unconstitutional taking of private property rights, and if challenged, these alternative roles will also mitigate the likelihood that government conduct will be judicially determined to be a taking of private property rights.

References

Bibliography

Ackerman, B. 1978. *Private Property and the Constitution.* New Haven, CT: Yale University Press.

Amar, A. 1998. *The Bill of Rights: Creation and Reconstruction.* New Haven, CT: Yale University Press.

Anderson, T., and P. Hill. 2003. The evolution of property rights. In *Property Rights: Cooperation, Conflict, and Law,* edited by T. Anderson and F. McChesney. Princeton, NJ: Princeton University Press.

Archer, J. 1994. *The Public Trust Doctrine and the Management of America's Coasts.* Boston, MA: University of Massachusetts Press.

Bentham, J. 1840. *Bentham's Theory of Legislation, Volume 1: Principles of Legislation.* Boston, MA: Weeks, Jordan, & Company.

Byrne, J. 2010. Rising seas and common-law baselines: A comment on regulatory takings discourse concerning climate change. *Vermont Journal of Environmental Law*, 11: 625.

Caldwell, M., and C. Segall. 2007. No day at the beach: Sea level rise, ecosystem loss, and public access along the California coast. *Ecology Law Quarterly*, 34: 533.

Craig, R. 2007. A comparative guide to the eastern public trust doctrines: Classification of states, property rights, and state summaries. *Penn State Environmental Law Review*, 16(1): 1.

Craig, R. 2009. *The Clean Water Act and the Constitution: Legal Structure and the Public's Right to a Clean and Healthy Environment, 2nd ed.* Washington, D.C.: Island Press.

Craig, R. 2010. A comparative guide to the western states' public trust doctrines: Public values, private rights, and the evolution toward an ecological public trust. *Ecology Law Quarterly*, 37(1): 53.

Davis. M. 1993. Lucas and takings: Private property redefined. *Kansas Journal of Law and Public Policy*, 2: 83.

Ely, J. 2008. Economic liberties and the original meaning of the Constitution. *San Diego Law Review*, 45: 673.

Fischel, W. 2003. Public goods and property rights: Of Coase, Tiebout, and just compensation. In *Property Rights: Cooperation, Conflict, and Law*, edited by T. Anderson and F. McChesney. Princeton, NJ: Princeton University Press.

Fisher, E. 2007. *Risk Regulation and Administrative Constitutionalism.* Oxford, UK: Hart Publishing.

Fisher, W. 1993. Trouble with Lucas. *Stanford Law Review*, 45: 1393.

Freyfogle, E. 2006. Goodbye to the public-private divide. *Environmental Law*, 36: 7, 11.

Freyfogle, E. 2007. *On Private Property: Finding Common Ground on the Ownership of Land.* Boston, MA: Beacon Press.

Holmes, O. 1909. *The Common Law.* Boston, MA: Little, Brown, and Co.

Intergovernmental Panel on Climate Change. 2007. IPCC Fourth Assessment Report: Climate Change 2007 (AR4). http://www.ipcc.ch/publications_and_data/publications_and_data_reports.shtml (accessed August 13, 2012).

Jagger, T., J. Elsner, and M. Saunders. 2008. Forecasting US insured hurricane losses. In *Climate Change and Society*, edited by H. Diaz and R. Murnane. New York: Cambridge University Press.

Kalo, J., R. Hildreth, A. Rieser, et al. 2007. *Coastal and Ocean Law: Cases and Materials.* St. Paul, MN: Thompson/West.

Killenbeck, M. (Ed.). 2002. *The Tenth Amendment and State Sovereignty: Constitutional History and Contemporary Issues.* Oxford, UK: Rowman & Littlefield.

Lazarus, R. 1993. Putting the correct 'spin' on Lucas. *Stanford Law Review*, 45: 1411.

Lazarus, R. 2008. Lucas unspun. *Southeastern Environmental Law Journal*, 16: 13.

Leuchtenberg, W. 2002. The Tenth Amendment over two centuries: More than a truism. In *The Tenth Amendment and State Sovereignty: Constitutional History and Contemporary Issues*, edited by M. Killenbeck. Oxford, UK: Rowman & Littlefield.

McGuire, C., and J. Hill. 2012. Climate adaptation and the Fifth Amendment of the U.S. Constitution: A regulatory takings analysis of adaptation strategies in coastal development with application to Connecticut's coastal management regime. *Sea Grant Law & Policy Journal*, 5(1): 140–168.

Repetto, R. (Ed). 2006. *Punctuated Equilibrium and the Dynamics of U.S. Environmental Policy.* New Haven, CT: Yale University Press.

Rose, C. 1999. Canons of property talk, or, Blackstone's anxiety. *Yale Law Journal*, 108: 601.

Rosenbloom, D., R. Kravchuk, and R. Clerkin 2008. *Public Administration: Understanding Management, Politics, and Law in the Public Sector, 7th ed.* New York: McGraw-Hill.

Sax, J. 2010. Some unorthodox thoughts about rising sea levels, beach erosion, and property rights. *Vermont Journal of Environmental Law*, 11: 641.

Schumann, A. (Ed). 2011. *Flood Risk Assessment and Management: How to Specify Hydrological Loads, Their Consequences and Uncertainties.* New York: Springer.

Titus, J. 1998. Rising seas, coastal erosion, and the takings clause: How to save wetlands and beaches without hurting property owners. *Maryland Law Review*, 57: 1279–1399.

Titus, J., D. Hudgens, D. Trescott, et al. 2009. State and local governments plan for development of most land vulnerable to rising sea level along the U.S. Atlantic coast. *Environmental Research Letters*, 4(4): 1.

U.S. Global Change Research Program. 2009. Global Climate Change Impacts in the United States (18). http://www.globalchange.gov/what-we-do/assessment/previous-assessments/global-climate-change-impacts-in-the-us-2009 (accessed August 13, 2012).

Ward, L. 2010. *John Locke and Modern Life.* Cambridge (UK) and New York: Cambridge University Press.

Table of Cases

Board of Regents v. Roth, 408 U.S. 564, 577 (1972).

Euclid v. Ambler Realty Company, 272 U.S. 365 (1926).

Grundy v. Thurston County, 117 P.3d 1089 (Wash. 2005).

Illinois Central Railroad v. Illinois, 146 U.S. 387 (1892).

Loretto v. Teleprompter Manhattan CATV Corporation, 458 U.S. 419 (1982).

Lucas v. South Carolina Coastal Council, 505 U.S. 1003 (1992).

Pennsylvania Central Transportation Company v. New York City, 438 U.S. 104 (1978).

Pennsylvania Coal v. Mahon, 260 U.S. 393 (1922).

Phillips v. Washington Legal Foundation, 524 U.S. 156, 164 (1998).

Severance v. Patterson (No. 09-0387), ___ S.W.3d ___ (Tex. 2012).

Shell Island Homeowners Association v. Tomlinson, 517 S.E.2d 401 (N.C. Ct. App. 1999).

Stop the Beach Renourishment Inc. v. Florida Department of Environmental Protection, 130 S.Ct. 2592 (2010).

Index

D

E

T

U